T0142903

Springer Theses

Recognizing Outstanding Ph.D. Research

Aims and Scope

The series "Springer Theses" brings together a selection of the very best Ph.D. theses from around the world and across the physical sciences. Nominated and endorsed by two recognized specialists, each published volume has been selected for its scientific excellence and the high impact of its contents for the pertinent field of research. For greater accessibility to non-specialists, the published versions include an extended introduction, as well as a foreword by the student's supervisor explaining the special relevance of the work for the field. As a whole, the series will provide a valuable resource both for newcomers to the research fields described, and for other scientists seeking detailed background information on special questions. Finally, it provides an accredited documentation of the valuable contributions made by today's younger generation of scientists.

Theses are accepted into the series by invited nomination only and must fulfill all of the following criteria

- They must be written in good English.
- The topic should fall within the confines of Chemistry, Physics, Earth Sciences, Engineering and related interdisciplinary fields such as Materials, Nanoscience, Chemical Engineering, Complex Systems and Biophysics.
- The work reported in the thesis must represent a significant scientific advance.
- If the thesis includes previously published material, permission to reproduce this must be gained from the respective copyright holder.
- They must have been examined and passed during the 12 months prior to nomination.
- Each thesis should include a foreword by the supervisor outlining the significance of its content.
- The theses should have a clearly defined structure including an introduction accessible to scientists not expert in that particular field.

More information about this series at http://www.springer.com/series/8790

Sébastien Andrieux

Monodisperse Highly Ordered and Polydisperse Biobased Solid Foams

Doctoral Thesis accepted by
University of Stuttgart, Stuttgart, Germany

 Springer

Author
Dr. Sébastien Andrieux
Institut Charles Sadron (UPR22—CNRS)
Strasbourg, France

Supervisor
Prof. Cosima Stubenrauch
Institute of Physical Chemistry
University of Stuttgart
Stuttgart, Germany

ISSN 2190-5053   ISSN 2190-5061   (electronic)
Springer Theses
ISBN 978-3-030-27834-2   ISBN 978-3-030-27832-8   (eBook)
https://doi.org/10.1007/978-3-030-27832-8

This Springer imprint is published by the registered company Springer Nature Switzerland AG
The registered company address is: Gewerbestrasse 11, 6330 Cham, Switzerland

À Alexis Dulaurent

Supervisor's Foreword

Many properties of solid foams depend on the distribution of the pore sizes and their organisation in space. However, these two parameters are very difficult to control in traditional foaming techniques. The Ph.D. thesis of Sébastien Andrieux is dealing with this difficulty and he was able to show that microfluidics can be used to tune the polydispersity of the foams (mono- vs different polydispersities) and the spatial organisation of the pores (ordered vs disordered). For this purpose, the microfluidic flow-focussing technique was modified such that the gas pressure oscillates periodically which translates into periodically oscillating bubble sizes in the liquid foam template. The liquid foams were generated from Chitosan solutions and then gelled via cross-linking with genipin before being freeze-dried to obtain a foamed aerogel with a specific structure. The study at hand fills two scientific gaps. On the one hand, a novel approach for the generation of foams with controlled polydispersity is presented. On the other hand, a solid foam consisting of rhombic dodecahedra is obtained. This structure has not yet been observed if one synthesises solid foams via foam templating.

The controlled variation of the foam's structure will allow studying systematically structure-property relations. Sébastien Andrieux's important contribution to the field was to establish a protocol for the synthesis of polydisperse Chitosan-based solid foams with a tuneable polydispersity, which will have an enormous impact on future work in this area. The protocol lends itself to studying the influence of the pore-size distribution on the mechanical properties of solid foams and is thus a promising tool for gaining a deeper understanding of structure-property relations of solid foams. Future work should explore a wider range of polydispersities and compare polydisperse foams with different polydispersities but the same average pore size. Moreover, being fully bio-based, this type of aerogel foams is a suitable candidate for applications in tissue engineering: open-cell, bio-based scaffolds with controllable pore sizes and pore-size distributions are heavily needed as they are believed to be extraordinary scaffolds for homogeneous and efficient cell growth.

Stuttgart, Germany
July 2018

Prof. Cosima Stubenrauch

Abstract

The aim of this work was the synthesis of monodisperse highly ordered bio-based polymer foams and a comparison with their polydisperse counterparts. We used the bio-based and biodegradable polymer chitosan, which we cross-linked with genipin. The polymer foams were synthesised via foam templating, i.e. via a liquid foam whose continuous phase contains a polymer and can be solidified. In order to obtain monodisperse highly ordered polymer foams, one first has to generate monodisperse highly ordered liquid foam templates. We did so by using microfluidics, which allows to produce monodisperse liquid foams with bubble sizes from 200 to 800 μm and polydispersities below 5%. The monodisperse foams were collected outside of the microfluidic channels and left to self-order under the influence of gravity and confinement.

We studied the kinetics of the cross-linking reaction to find the optimal storage conditions during cross-linking. Once cross-linked, we freeze-dried the gelled foams to obtain solid chitosan foams. We compared the morphological properties of the solid foams with those of the liquid templates in order to test the efficiency of the developed templating route. We observed how modifying the cross-linking and drying conditions can strongly affect the morphology of the solid foams. The main issue was to maintain the key properties of the liquid foam template throughout the solidification process, namely the bubble size distribution, the structural order and the density.

We then compared the synthesised monodisperse polymer foams with their polydisperse counterparts. Although easy foaming methods exist for the generation of polydisperse foams, they do not allow the control over the polydispersity. We thus used microfluidics to generate liquid chitosan foams with tunable polydispersities from below 5% up to 26%. Microfluidics allows to match the average bubble size and density of the polydisperse liquid chitosan foam with those of the monodisperse counterpart. After solidifying the liquid templates, we obtained solid foams with controlled polydispersities and studied the influence of the polydispersity on the mechanical properties. However, we observed that not the polydispersity but the foam density was the main parameter at play. Moreover, the solid chitosan foams had weak mechanical properties with elastic moduli below 100 kPa.

To overcome this issue, we incorporated cellulose nanofibres to the original chitosan solution and followed the developed route for foam templating. We had to adapt the microfluidic parameters to account for the viscosity changes brought about by the nanofibres. However, we managed to produce monodisperse liquid foams having the same bubble size, i.e. ~ 300 μm, but different amounts of cellulose nanofibres. The cellulose content had a strong influence on the solid foam morphology in general and on the pore connectivity in particular.

Parts of this thesis have been published in the following journal articles:

S. Andrieux, W. Drenckhan, C. Stubenrauch, Highly ordered monodipserse scaffolds: from liquid to solid foams. *Polymer* **2017**, *126*, 425–431.

S. Andrieux, W. Drenckhan, C. Stubenrauch, Generation of solid foams with controlled polydispersities using microfluidics. *Langmuir* **2018**, *34*, 1581–1590.

S. Andrieux, A. Quell, C. Stubenrauch, W. Drenckhan, Liquid foam templating—a route to tailor-made polymer foams *Adv. Colloid Interface Sci.* **2018**, *256*, 276–290.

Acknowledgements

I would like to thank first and foremost Prof. Dr. Cosima Stubenrauch for giving me the opportunity to do my Ph.D. Thesis in her group, as well as for the exceptional supervision and mentoring that followed. I am thankful for her philosophy as regards scientific research and how open to new ideas and collaborations she is. I am grateful for the investments she made in my project and myself by sending me to many international conferences and by helping me to develop my network.

I am grateful to Prof. Dr. Dominique Langevin for accepting to be the second referee for my thesis and for hearing my defence. I thank likewise Prof. Dr. Sabine Laschat for accepting to be the chairwoman during my defence.

I warmly thank Dr. Wiebke Drenckhan for her help and mentoring from the very beginning of my thesis. Every discussion with her lead to a deeper understanding of foams or new ideas.

I am very grateful to Prof. Dr. Alexander Fels for his help with the SEM measurements and his precious advice. I would also like to thank Sven Richter for the use of the freeze-drier. I am grateful to the MMOI team at the LPS and particularly to Clément Honorez and to Dr. Anaïs Giustiniani for their help with the design of the microfluidic chips. I would also like to thank Dr. Angelika Menner and her team at the University of Vienna for welcoming me to the group and for their help with the compression tests and density measurements. I also thank Dr. Thierry Roland and his colleagues at the ICS for the many fruitful discussions about the mechanics of my foams. I would also like to warmly thank Lilian Medina from KTH Stockholm, who showed enough interest in my project to start a constructive collaboration. I also thank his supervisor Prof. Dr. Lars Berglund for allowing this collaboration to happen and for his useful advice. A million thanks go as well to Prof. Dr. Andrea Barbetta and Dr. Marco Costantini for stimulating discussions and for performing the μCT scans. I also thank Raouf Jemmali for his input on μCT.

I would like to thank and congratulate my students, Tamara Schad, Carina Schellenberg, Matthias Hermann, Anastasia Tsianaka and Michael Herbst, who helped on my project.

I am very grateful to Diana Zauser and Birgit Feucht for their support regarding the laboratories. I am also thankful to the staff from the electronic, mechanical and glass workshops, who helped give life to many ideas. More generally, I would like to thank the past and present co-workers of the Stubenrauch group for creating a lovely work atmosphere and a philosophy of mutual assistance within the group.

A huge thank you goes also to all the people close to me, my family and friends, who helped make my time in Germany a great personal adventure.

Contents

Nomenclature

Abbreviations

μCT	Micro-computed tomography
AcOH	Acetic acid
cac	Critical aggregation concentration
cmc	Critical micellar concentration
CNF	Cellulose nanofibre
COC	Cyclic olefin copolymers
COC_{170}	COC with a glass transition temperature of 170 °C
COC_{80}	COC with a glass transition temperature of 80 °C
DD	Deacetylation degree
DNA	Deoxyribonucleic acid
DP	Degree of polymerisation
e-CNF	Enzymatic cellulose nanofibre
EtOH	Ethanol
GLY	Glycerol
GPC	Gel permeation chromatography
HEMA	2-hydroxyethyl methacrylate
HIPE	High internal phase emulsion
IUPAC	International Union of Pure and Applied Chemistry
NaOAc	Sodium acetate
PDMS	Polydimethylsiloxane
PEG	Polyethylene glycol
polyHIPE	Polymerised high internal phase emulsion
PPG	Polypropylene glycol
PVA	Polyvinyl alcohol
RT	Room temperature
SEM	Scanning Electron Microscopy

Constants

N_A Avogadro number = $6.022 \ 10^{23} \ \text{mol}^{-1}$
R Gas constant = $8.314 \ \text{J} \ \text{K}^{-1} \ \text{mol}^{-1}$
g   Gravitational acceleration = $9.81 \ \text{m} \ \text{s}^{-2}$

Parameters

$\dot{\gamma}$	Shear rate
$\bar{d}$	Average bubble/pore diameter
ϕ	Volume fraction of polymer distributed in the struts
$\tilde{h}$	Reduced foam height
α	Dimensionless exponent
Δl	Height difference before and after compression
η	Dynamic viscosity
γ	Surface tension
Π_d	Disjoining pressure
Π_o	Osmotic pressure
ρ^*	Relative density
ρ_{foam}	Foam density
ρ_{polymer}	Polymer density
ρ	Density
σ	Stress
σ_r	Stress at rupture
σ_y	Yield stress
υ	Average velocity of a fluid particle
τ	Period
ε	Strain
φ	Liquid fraction
φ^*	Critical liquid fraction for the energy density
$\vec{v}_x$	Velocity of a fluid particle along the x axis
A	Interfacial area
A_c	Square section of the chip channel
b_0	Monomer length
Bo	Bond number
c_i	Concentration of the species i
Ca	Capillary number
D	Characteristic dimension of the system
d	Bubble/pore diameter
d_{cc}	Centre-to-centre distance
D_c	Length of the square section of the chip's channel
E	Elastic modulus (Young's modulus)

E_A	Activation energy
E_s	Surface energy
F	Force
f	Gel fraction
f_b	Bubbling frequency
f_c	Critical gel fraction
G'	Storage modulus
G''	Loss modulus
h_0	Initial height of the foam
h_c	Height of the channel
h_{film}	Thickness of the foam film
h_t	Height of the foam at time t
L	Edge length
l	Height of the sample
l_0	Initial height of the sample
l_c	Capillary length
l_{PB}	Length of the Plateau border
$M_w(x)$	Molecular weight of the molecule or monomer x
N	Number of bubbles/pores of diameter d
N_{total}	Total number of bubbles/pores available for counting
P	Porosity
p_0	Atmospheric pressure
p_{chit}	Pressure applied on the chitosan solution
p_c	Capillary pressure
$p_{gas,max}$	Maximum gas pressure
$p_{gas,min}$	Minimum gas pressure
p_{gas}	Gas pressure of the microfluidic set-up
p_g	Gas pressure in a bubble
p_h	Hydrostatic pressure
p_L	Laplace pressure
p_l	Pressure of the liquid in a foam
Q	Flow rate
Q_g	Gas flow rate
Q_l	Liquid flow rate
$Q_{chitosan}$	Flow rate of the chitosan solution
$Q_{genipin}$	Flow rate of the genipin solution
R	Bubble radius
r	Radius of curvature of the Plateau border
R_{32}	Sauter mean radius
R_o	Characteristic dimension of the orifice/channel
r_{PB}	Thickness of the Plateau border
Re	Reynolds number
S	Surface area of the sample
T_g	Glass transition temperature

t_{FD}	Freeze-drying time
T_{gel}	Gelation temperature
t_{gel}	Gelation time
τ_d	Characteristic drainage time
V_b	Bubble volume
V_o	Critical blocking volume for the squeezing regime
V_{PB}	Volume of the Plateau border
v_x	x-component of the velocity of a fluid particle
w_c	Width of the chip's channel
We	Weber number

List of Figures

List of Tables

Chapter 1
Introduction

1.1 Motivation

A foam is a dispersion of gas in a continuous phase. If the continuous phase is liquid (solid), the system is a liquid (solid) foam. A liquid foam is thermodynamically unstable; its structure evolves with time. Despite a relatively short lifespan, we find liquid foams in many daily-life products. They have been thoroughly studied for applications in, for example, the food [18] and cosmetic industries [2]. Solid foams are obtained by solidifying liquid foams, with a more or less good retention of the structure during solidification [19, 20, 22]. Liquid foams may lead to open-cell foams (sponge-like materials) or closed-cell foams, depending on whether or not the film separating two bubbles resists solidification (see Fig. 1.1).

Solid foams are important due to their mechanical, insulating, and shock absorption properties as well as due to their low weight [10]. They are used in a variety of applications ranging from house insulation to packaging. The Thesis at hand focuses on solid foams consisting of polymeric materials. Metal or ceramic foams will not be discussed in this work, i.e. the solid foams we refer to are polymer foams or (macro)porous polymers. Polymer foams can be produced with the help of various techniques such as foam injection moulding, foam extrusion or thermoset reactive foaming, and can be made of many different materials such as polystyrene, polyurethane, polyolefins or starch [20]. Because they can be formulated and processed to have specific chemical and physical properties, they have a wide range of applications. Nonetheless, all the production processes have the same drawback: the lack of a satisfying control over the pore size distribution and over the pore connectivity (whether the foam has an open-cell, closed-cell or intermediate structure). Indeed, the morphology of a polymer foam needs to be well controlled in order to have the properties required by the application aimed for. While, for example, thermal insulation or flotation call for a closed-cell structure, absorbent materials and scaffolds for tissue engineering call for open-cell foams. Moreover, the mechanical properties of polymer foams do not only depend on the mechanical properties of the

© Springer Nature Switzerland AG 2019   1
S. Andrieux, *Monodisperse Highly Ordered and Polydisperse Biobased Solid Foams*,
Springer Theses, https://doi.org/10.1007/978-3-030-27832-8_1

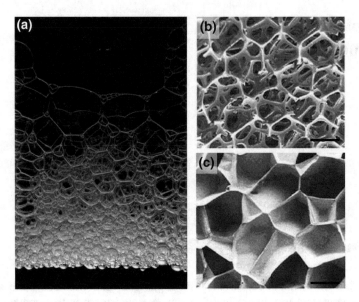

Fig. 1.1 Examples of **a** a liquid foam (© Wiebke Drenckhan), **b** an open-cell [19] and **c** a closed-cell solid foam [19]. The scale bars are 500 μm

respective bulk polymer but also on their density and structure. Hence the importance of being able to control these parameters while generating macroporous polymers. Besides, current production processes require high temperatures and pressures. It is therefore imperative for environmental and economic reasons to significantly reduce the amount of energy required to produce such materials.

Academic researchers have turned their attention to emulsion templating as a way to better control the structure of polymer foams. In emulsion templating one first generates a high internal phase emulsion (referred to as HIPE, which is an emulsion that contains at least 74 vol% dispersed phase) using a monomer as the continuous phase [8, 9, 21, 30]. One obtains a polymer foam by polymerising the continuous phase and extracting the dispersed phase. Materials synthesised via this method are named polyHIPEs, which stands for polymerised high internal phase emulsions. A similar approach called foam templating was later developed. This technique is akin to emulsion templating, except that the dispersed phase is not a liquid but a gas [1, 4, 6, 26]. However, one typically makes liquid foams from an aqueous solution in the presence of surfactant to stabilise the foam. Foam templating thus calls for a monomer which can be foamed upon addition of surfactant, i.e. a polar monomer in which the surfactant is soluble. The monomer foam template has to remain stable during polymerisation. One can also generate a liquid foam template by using a water-soluble polymer, e.g. a polysaccharide, dissolved in an aqueous solution [3–5, 13]. However, Murakami and Bismarck [26] managed to synthesise solid foams from hydrophobic monomers by using particles with a well-suited oleophobicity to generate a surfactant-free particle-stabilised foam template.

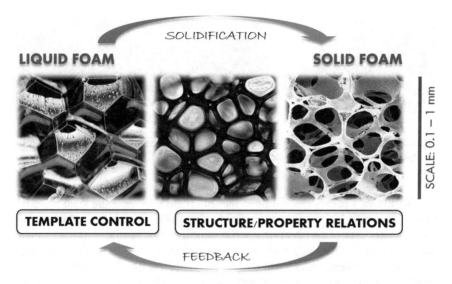

Fig. 1.2 General concept of foam templating showing how the structure of the liquid foam and the solidification procedure need to be controlled to control the structure of the solid foam, and thus its properties. Taken from [1]

Yet, foam templating fails to produce well-defined structures. Since the structure of the liquid template dictates the structure of the solid foam, one first needs to generate tailor-made liquid templates in order to generate tailor-made solid foams. Foam templating thus requires consistent feedbacks to find out how to modify the way the foam template is either formed or solidified to obtain the properties aimed for (see Fig. 1.2).

The chase for well-defined liquid foams motivated many studies on monodisperse liquid foams. Monodisperse liquid foams, in which all bubbles have the same size, possess interesting properties as they can self-order into well-defined periodical structures. In other words, they crystallise. Producing monodisperse polymer foams with a controllable structure from monodisperse liquid templates is of utmost interest from both a scientific and an engineering point of view. Monodisperse foams are fundamentally interesting as they are homogeneous throughout their whole volume, which allows scientists to study the influence of the different morphological parameters (i.e. the pore size, the density, the size of the interconnects) on the properties of the foams, as the structure/properties relationships are a foremost topic of interest in the polymer foam community. Such homogeneous materials are also sought for in practical domains such as biomedicine, which aims at developing the perfect scaffold for cell growth and tissue engineering. It is a consensus that the ideal scaffold for tissue engineering has to (a) be monodisperse and homogeneous (i.e. highly ordered), and (b) have pore sizes and interconnect sizes which are adapted to the cells one needs to grow [5, 6, 12]. The demand for monodisperse highly ordered polymer foams is thus strong.

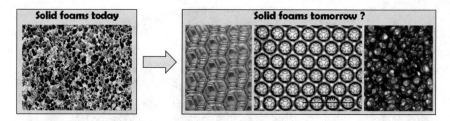

Fig. 1.3 Goal of the foam templating community which uses microfluidics as a tool to generate the liquid foam template. Adapted from [38]

Microfluidics constitutes the most feasible route to developing such materials. Indeed, Lab-on-a-Chip devices allow for the formation of bubbles one by one, with such a level of control over their sizes and volume fractions that the resulting foams are monodisperse. To date, only a few studies have addressed the synthesis of monodisperse polymer foams using foam templating [4, 6, 14, 16, 31, 35–37, 39], but a small community has developed over the years to aim for structures such as the ones shown in Fig. 1.3.

Today's challenges focus on a green and responsible chemistry. This also applies to materials science, where the search for biobased and/or biodegradable materials is highly active. Biodegradable polymers are also of interest for applications in biology and medicine, as biopolymers may show biocompatibility as well: they are not rejected by living bodies. A good example is tissue engineering, for which some research groups have already developed such materials via emulsion templating [11, 17, 25, 32], but also foam templating [4, 6, 7, 12, 13, 15, 16]. The Thesis at hand aims at continuing the work done in the field of monodisperse polymer foams towards a greener chemistry while providing a better understanding of the structure-properties relationships. Some steps have already been taken in this direction [12, 23, 36] by using cross-linking biopolymers to produce hydrogels (see Fig. 1.4).

1.2 Task Description

This Thesis aims at synthesising monodisperse biobased polymer foams. The polymer used in this project is chitosan. Chitosan is a polysaccharide obtained from the deacetylation of chitin, which is a biodegradable biopolymer present in the exoskeleton of crustaceans and in mushrooms [27]. It can also be produced from the fermentation of some species of fungi and yeasts [29]. Chitosan is slightly soluble in dilute acidic solutions [34] and forms a hydrogel upon the addition of cross-linker [28, 33]. In other words, chitosan solutions are "green" and can be gelled via the cross-linking of chitosan. An appropriate surfactant for foam stability and a cross-linker are thus required, with the constraint of having to be biobased as well. We chose an

Fig. 1.4 Photographs of ordered wet (left) and dry (right) chitosan foams obtained by microfluidic techniques (from [36])

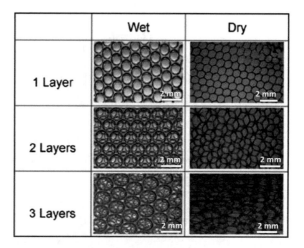

alkyl polyglycoside–a biobased sugar surfactant–and genipin, respectively, the latter being a natural molecule extracted from gardenia.

Miras et al. were able to produce macroporous chitosan cross-linked with genipin via emulsion templating [23, 24]. Although the HIPEs were obtained by drop break-up under shear, and were thus not monodisperse, this work can be considered as a "proof of concept" for the synthesis of solid chitosan foams from a liquid template. However, the system used by Miras et al. was not fully biobased, as the surfactant used was a synthetic alcohol ethoxylate (Synperonic A7) and the dispersed oil phase was n-decane.

Testouri et al. presented some preliminary studies on monodisperse chitosan solid foams [36] (see Fig. 1.4) with pores much larger than those obtained by Miras et al. [23, 24], namely 1–3 mm versus 1–10 μm. However, the system was not entirely biobased either— neither the surfactant nor the cross-linker was biobased. Moreover, the mechanical properties were not studied, the effects of changes in the formulation of the system were not investigated, and the pore size was not systematically varied. The present work is intended to optimise this system by solely using biobased compounds and to provide a detailed study of the generation protocol and of the final properties of the solid foams. For this purpose, foam templating using microfluidics is used to produce fully biobased monodisperse chitosan foams. Microfluidics also gives access to a wide range of bubble sizes, allowing us to explore pore sizes between 50 μm and 1 mm. The general concept of producing monodisperse chitosan liquid foams via microfluidics is presented in Fig. 1.5.

Times scales are of tremendous importance in foam templating. The monodisperse liquid template needs to remain stable during gelation so that the solid foam resembles its liquid foam template. Once the chitosan is cross-linked, it forms a hydrogel, and the foam is no longer subject to destabilising mechanisms such as Ostwald ripening or coalescence (see Sect. 2.1.2). One thus aims at reducing the gelation time to a minimum. The microfluidic set-up is not only necessary for the formation of

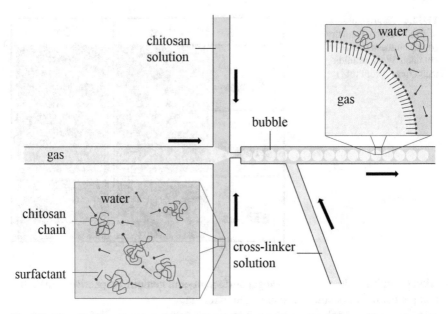

Fig. 1.5 Microfluidic-assisted production of monodisperse chitosan liquid foams. The arrows indicate the directions of the different flows

monodisperse bubbles, but it also facilitates the addition of the cross-linker required for the chitosan to gel. Preliminary studies are therefore required to determine the right formulation, so that the chitosan will not gel too early (and as a result clog the chip channels), or too late. One thus has to find the optimal gelation time for which the foam sample does not have enough time to destabilise, but the polymer solution does not solidify in the microfluidic set-up. Since the polymer is dissolved in water, one also needs to dry the foam template to obtain a dry, solid polymer foam. The drying step is of tremendous importance because the challenge is to remove a large part of the material without hurting its structure. We will thus investigate how drying may affect the morphology of the resulting solid foams.

A frequently asked question when discussing monodisperse solid foams is: "What are their advantages over polydisperse foams?" Indeed, even if monodispersity and order are required for some specific applications such as tissue engineering, we still lack experimental proof of the possible improvements of, e.g., the mechanical or thermal properties of the polymer foams that may come along with monodispersity. This work has the ambition to provide a starting point for such studies and launch the debate in the polymer foam community from two different angles. Firstly, we want to show how to control the pore size distribution if one aims at studying its influence on the mechanical properties of solid foams. Secondly, we want to open the way to monodisperse composites in order to improve the mechanical properties of polysaccharide-based solid foams, which are per se mechanically weak, without having to sacrifice the monodispersity of the foams.

References

1. Andrieux S, Quell A, Drenckhan W, Stubenrauch C (2018) Adv Colloid Interface Sci 256:276–290
2. Arzhavitina A, Steckel H (2010) Int J Pharm 394(1–2):1–17
3. Barbetta A, Dentini M, De Vecchis M, Filippini P, Formisano G, Caiazza S (2005) Adv Funct Mater 15(1):118–124
4. Barbetta A, Gumiero A, Pecci R, Bedini R, Dentini M (2009) Biomacromolecules 10(12):3188–3192
5. Barbetta A, Barigelli E, Dentini M (2009) Biomacromolecules 10(8):2328–2337
6. Barbetta A, Rizzitelli G, Bedini R, Pecci R, Dentini M (2010) Soft Matter 6:1785–1792
7. Barbetta A, Carrino A, Costantini M, Dentini M (2010) Soft Matter 6:5213–5224
8. Bartl VH, Von Bonin W (1962) Die makromolekulare chemie 57(1):74–95
9. Cameron NR (2005) Polymer 46:1439–1449
10. Cantat I, Cohen-Addad S, Elias F, Graner F, Höhler R, Pitois O, Rouyer F, Saint-Jalmes A (2013) Foams - structure and dynamics. Oxford University Press, Oxford
11. Christopher GF, Anna SL (2007) J Phys D 40(19):R319
12. Chung K-Y, Mishra NC, Wang C-C, Lin F-H, Lin K-H (2009) Biomicrofluidics 3(2):022403
13. Colosi C, Costantini M, Barbetta A, Pecci R, Bedini R, Dentini M (2013) Langmuir 29(1):82–91
14. Costantini M, Colosi C, Guzowski J, Barbetta A, Jaroszewicz J, Święszkowski W, Dentini M, Garstecki P (2014) J Mater Chem B 2(16):2290–2300
15. Costantini M, Colosi C, Jaroszewicz J, Tosato A, Święszkowski W, Dentini M, Garstecki P, Barbetta A (2015) ACS Appl Mater Interfaces 7(42):23660–23671
16. Costantini M, Colosi C, Mozetic P, Jaroszewicz J, Tosato A, Rainer A, Trombetta M, Święszkowski W, Dentini M, Barbetta A (2016) Mater Sci Eng C 62:668–677
17. David D, Silverstein MS (2009) J Polym Sci A 47(21):5806–5814
18. Dickinson E, Lorient D (1995) Food macromolecules and colloids. RSC Publishing, Cambridge
19. Gibson LJ, Ashby MF (1997) Cellular solids: structure and properties (Cambridge solid state science series). Cambridge University Press, Cambridge
20. Lee S-T, Park CB, Ramesh NS (2006) Polymeric foams: science and technology. Taylor and Francis Inc, Bosa Roca
21. Lissant KJ (1974) Emulsions and emulsion technology. Taylor and Francis Inc, New York
22. Mills N (2007) Polymer foams handbook: engineering and biomechanics applications and design guide. Elsevier Science and Technology, Oxford
23. Miras J, Vílchez S, Solans C, Esquena J (2013) J Colloid Interface Sci 410:33–42
24. Miras J, Vílchez S, Solans C, Tadros T, Esquena J (2013) Soft Matter 9:8678–8686
25. Moglia RS, Holm JL, Sears NA, Wilson CJ, Harrison DM, Cosgriff-Hernandez E (2011) Biomacromolecules 12(10):3621–3628
26. Murakami R, Bismarck A (2010) Adv Funct Mater 20(5):732–737
27. Muzzarelli RA (1973) Natural chelating polymers; alginic acid, chitin and chitosan. Pergamon Press, Oxford
28. Nyström B, Kjøniksen A-L, Iversen C (1999) Adv Colloid Interface Sci 79:81–103
29. Pochanavanich P, Suntornsuk W (2002) Lett Appl Microbiol 35(1):17–21
30. Pulko I, Krajnc P (2012) Macromol. Rapid Commun 33(20):1731–1746
31. Quell A, Elsing J, Drenckhan W, Stubenrauch C (2015) Adv Eng Mater 17(5):604–609
32. Robinson JL, Moglia RS, Stuebben MC, McEnery MA, Cosgriff-Hernandez E (2014) Tissue Eng Part A 20(5–6):1103–1112
33. Rohindra DR, Nand AV, Khurma JR (2004) SPJNAS 22(1):32–35
34. Sorlier P, Denuzière A, Viton C, Domard A (2001) Biomacromolecules 2(3):765–772
35. Testouri A (2012) Highly structures polymer foams from liquid foam templates using millifluidic lab-on-a-chip techniques, Université Paris-Sud XI, Ph.D. Thesis
36. Testouri A, Honorez C, Barillec A, Langevin D, Drenckhan W (2010) Macromolecules 43(14):6166–6173

37. Testouri A, Arriaga L, Honorez C, Ranft M, Rodrigues J, van der Net A, Lecchi A, Salonen A, Rio E, Guillermic R-M, Langevin D, Drenckhan W (2012) Colloids Surf A 413:17–24
38. Testouri A, Ranft M, Honorez C, Kaabeche N, Ferbitz J, Freidank D, Drenckhan W (2013) Adv Eng Mater 15(11):1086–1098
39. van der Net A, Gryson A, Ranft M, Elias F, Stubenrauch C, Drenckhan W (2009) Colloids Surf A 346:5–10
40. Weaire D, Hutzler S, Cox S, Kern N, Alonso MD, Drenckhan W (2003) J Phys Condens Matter 15(1):S65–S73

Chapter 2
Theoretical Background

2.1 Liquid Foams

Foams are dispersions of gas bubbles in a continuous phase. This continuous phase can either be a liquid, then one speaks of liquid foams (such as the foam on top of a beer or the foam in the washing machine); or a solid, then one speaks of solid foams (such as Styrofoam or polyurethane foams used for house insulation). In this work, we are interested in both liquid and solid foams, but let us first talk about liquid foams.

2.1.1 Liquid Foams at all Length Scales

The stability and structure of liquid foams depend on complex parameters acting at different length scales but which are all interdependent. Let us consider the physics of a liquid foam through the scope of its different length scales, from the smallest to the largest, as sketched in Fig. 2.1.[1]

The gas-liquid interface The main feature of liquid foams is their large surface area, i.e. the interface between the gas phase and the liquid phase, as sketched in Fig. 2.1A. This surface area dictates the surface energy E_s of a given foam, such as

$$E_s = \gamma A, \tag{2.1}$$

---

[1] We do not discuss here the generation of foams and foamability. Although very interesting, this topic has not much relevance here since the work at hand deals with foams generated via microfluidics, to which we dedicate an entire section (Sect. 2.4). For extended information on foam generation, we recommend the reading of [40].

© Springer Nature Switzerland AG 2019
S. Andrieux, *Monodisperse Highly Ordered and Polydisperse Biobased Solid Foams*,
Springer Theses, https://doi.org/10.1007/978-3-030-27832-8_2

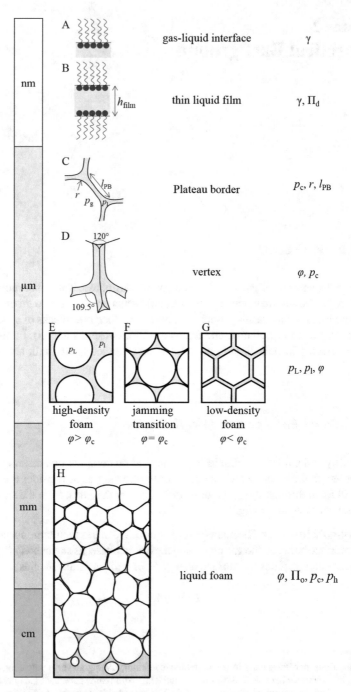

Fig. 2.1 Schematic representation of liquid foams at different length scales with the relevant physical parameters for each length scale (inspired from [19, 30])

where γ is the surface tension and A the interfacial area [46]. The surface energy is directly proportional to the area of the interface: creating twice more interface requires twice more energy. The system tends towards the minimisation of this surface energy by reducing the interfacial area A. The surface tension γ, which also dictates the surface energy, is defined as the energy required to produce an interface of a unit area. Its unit is thus $J\,m^{-2}$, which is often expressed as $mJ\,m^{-2}$ or $mN\,m^{-1}$ [42]. The surface tension can be decreased via the use of surfactants, which are literally "surface active agents". As shown in Fig. 2.2, surfactants are composed of a hydrophilic head which "likes" water and polar solvents, and a hydrophobic tail which "does not like" water and polar solvents and has a stronger affinity for apolar phases. This dual character is called amphiphilicity and translates into an equilibrium between free surfactants in the liquid phase and surfactants adsorbing at the gas-liquid interface in case of polar liquids [42].

At low surfactant concentrations $c_{surfactant}$, only a few surfactant molecules are adsorbed at the interface (Fig. 2.2), which does not suffice to induce a noticeable effect on the surface tension γ (Fig. 2.2). One measures thus a surface tension close to that of the pure liquid. As the surfactant concentration $c_{surfactant}$ increases, the interface covered with surfactants becomes larger and the surface tension γ decreases accordingly. Since the interface has a fixed area, the surfactant concentration reaches a point at which surfactant molecules cover the whole interface. From this surfactant concentration on the surface tension remains constant upon addition of surfactant. At even higher concentrations, the surfactants self-assemble into structures called micelles which are thermodynamically more favourable than dispersed free surfactant molecules. The concentration at which micelles start to build is called the critical micellar concentration (cmc) [28, 42].

Let us go back to considering the liquid-gas interface as part of a liquid foam: the foam strives to lower its surface energy E_s as much as possible, which practically consists in reaching for the lowest surface tension γ possible. The cmc of a surfactant is thus the first parameter to measure in order to place oneself above this concentration when generating a liquid foam. Given that the surface tension does not decrease further upon addition of surfactant above the cmc, one does not need to use too large amounts of surfactant to help stabilise the foam; the surfactant could otherwise act as a cosolvent.

The thin liquid film Thermodynamically speaking, it is not favourable for two isolated bubbles to come in contact and deform, as it increases the overall interfacial area A—the sphere being the geometric object with the lowest area for a given volume—and thus the surface energy E_s. To bring two bubbles in close contact to form a thin film, one needs to push them together by applying a constant force F. The bubbles will approach and deform, creating an interfacial thin film of thickness h_{film}, as shown in Fig. 2.1B. The film will get thinner down to the point where the disjoining pressure Π_d is high enough to counterbalance this force per newly created area. The disjoining pressure Π_d varies with the film thickness h_{film} and corresponds to how far both liquid-air interfaces "see" each other [108]. The disjoining pressure results from three different interactions: the repulsive electrostatic interactions, the repulsive

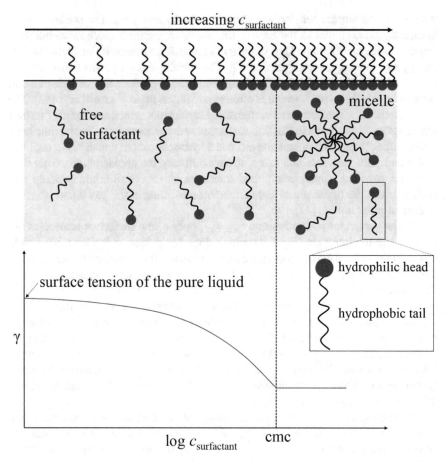

Fig. 2.2 Liquid-gas interface in presence of surfactant. (top) The increase of the surfactant concentration is schematically shown from left to right. (bottom) The graphic shows the variation of the surface tension γ with the surfactant concentration $c_{\text{surfactant}}$

steric interactions, and the attractive van der Waals interactions. The electrostatic and steric components favour film stabilisation whereas the van der Waals forces favour its rupture [42, 48, 108].

The Plateau border and vertex Once all the forces are in equilibrium, the disjoining pressure Π_d equals the capillary pressure p_c, which is equal to the difference between the gas pressure p_g and the liquid pressure p_l, i.e. $p_c = p_g - p_l$, as sketched in Fig. 2.1C. A three-dimensional representation of a Plateau border is shown in Fig. 2.3.

Three rules called the Plateau rules set the structure of a foam and foam films: (i) the foam films have a constant radius of curvature which is set by Laplace's law (Eq. 2.2), (ii) the films always meet in the number of three to form a Plateau border with an angle of 120°, (iii) the Plateau borders always meet in the number of four

Fig. 2.3 Structure of a
Plateau border showing the
different radii at play. R is
the radius of curvature of the
bubble and r_{PB} is the
thickness of the Plateau
border, which can be
considered as equal to the
radius of curvature of the
Plateau border r (redrawn
from [19])

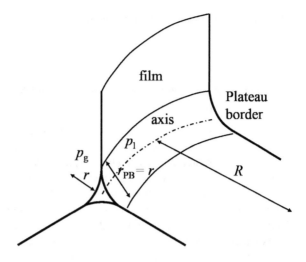

to form a vertex, with an angle of $\sim 109.5°$ [121]. The Laplace's law which sets the
first Plateau rule reads

$$p_c = p_g - p_l = \gamma \left(\frac{1}{r} + \frac{1}{R} \right), \tag{2.2}$$

with R being the radius of curvature of the Plateau border along its axis—which can
be assimilated to the radius of curvature of the bubble—as shown in Fig. 2.3, and r
being the radius of curvature of the Plateau border [19]. For foams with low liquid
fractions φ such as sketched in Figs. 2.1C, D and 2.3, $r \ll R$ and Eq. 2.2 becomes

$$p_c = p_g - p_l \simeq \frac{\gamma}{r}. \tag{2.3}$$

If one increases the liquid fraction φ by adding a liquid which fills the Plateau borders,
the radius of curvature of the Plateau borders r increases and the capillary pressure
p_c decreases. Let us have a closer look at the liquid fraction φ. It is the amount of
liquid in the foam. We have already discussed that the foam films are very thin when
all forces are at equilibrium. We can thus assume that most of the liquid is contained
in the Plateau borders. One can thus determine the liquid fraction by looking at a
single bubble. The volume of the foam is then scaled down to the bubble volume V_b,
which one approximates to R^3. The volume of the Plateau border V_{PB} is equal to its
cross-section—which one approximates to r^2—times its length l_{PB}. Assuming that
the length of the Plateau border l_{PB} is of the order of the bubble radius R, one can
write

$$\varphi \simeq \frac{V_{PB}}{V_{bubble}} \simeq \frac{r^2 R}{R^3}, \tag{2.4}$$

which leads to

$$r \simeq \sqrt{\varphi} R. \tag{2.5}$$

The capillary pressure can be thus rewritten as follows [19]

$$p_c = p_l - p_g \simeq \frac{\gamma}{\sqrt{\varphi}R}. \tag{2.6}$$

Thus, the lower the liquid fraction φ is—or the smaller the bubbles are—, the higher is the capillary pressure p_c.

Liquid foam As shown in Fig. 2.1E–G, the shape of the bubbles is highly dependent on the liquid fraction φ. Above a critical liquid fraction φ_c the bubbles are spherical and are not systematically in close contact. One speaks of a high-density foam. At the critical liquid fraction, i.e. $\varphi = \varphi_c$, the spherical bubbles are in close contact. Remove liquid to reach a liquid fraction lower than the critical liquid fraction and the bubbles deform to become polyhedral, resulting in a so-called low-density foam.[2] The transition from spherical bubbles to polyhedral bubbles at the critical liquid fraction is called the jamming transition. One can also consider the transition from polyhedral bubbles to spherical bubbles upon addition of liquid, the rigidity-loss transition, which also occurs at the same critical liquid fraction φ_c. Note that the physics of packing dictates the critical liquid fraction. Therefore, $\varphi_c = 0.26$ for close-packed monodisperse spheres, whereas $\varphi_c = 0.36$ for randomly packed monodisperse spheres, also known as Bernal's packing, [5]. The case of monodisperse foams is specific (see Sect. 2.1.3) and foams are more often polydisperse than not. Note that the critical liquid fraction decreases when the polydispersity increases, as the smaller bubbles can fill the voids between the larger bubbles.

Foams are submitted to gravity and the liquid drains out of the foam films (see Sect. 2.1.2) leading to a liquid fraction gradient with the foam height which reads [39]

$$\varphi(\tilde{h}) = \frac{\varphi_c}{(1 + \tilde{h})^2}. \tag{2.7}$$

One uses the reduced foam height $\tilde{h} = h \cdot R_{32}/l_c^2$ to include the different parameters of the liquid foam in the liquid fraction profile presented in Eq. 2.7. R_{32} is the Sauter mean radius defined as $<R^3> / <R^2>$ and l_c is the capillary length defined as $l_c = \sqrt{\gamma/\rho g}$, with g the gravitational acceleration and ρ the density of the liquid. As a result, Eq. 2.7 holds for all systems, whatever the surface tension or the bubble size. One can, however, neglect the action of gravity if the hydrostatic pressure p_h is much smaller than the capillary pressure p_c. One can thus write that if

$$\rho g h \ll \frac{\gamma}{\sqrt{\varphi}R}, \tag{2.8}$$

---

[2]The liquid foam literature usually uses the terms "wet foams" and "dry foams" for high-density foams and low-density foams, respectively. We chose to avoid using the terminology "wet" and "dry" since the present work deals with liquid foams which are dried. We will thus not talk about "wet foams" and use the adjective "dry" in its primary meaning, i.e. not containing water, to not confuse the reader.

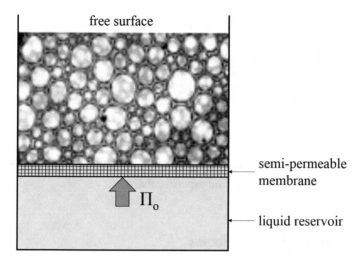

Fig. 2.4 Experiment showing the macroscopic action of the osmotic pressure Π_o on a liquid foam. Figure redrawn from [60]

one can assume that the gravity is negligible below a height h [71]

$$h \ll \frac{l_c^2}{\sqrt{\varphi}R}. \tag{2.9}$$

In other words, the foam is considered homogeneous below this foam height, and the liquid fraction is regarded as constant. The liquid fraction profile comes from the interplay of different forces. Although gravitation tends to empty the Plateau borders and reduce the liquid fraction, we have seen that below the critical liquid fraction φ_c, reducing the liquid fraction results in a deformation of the bubbles and the creation of interfacial area, which costs energy (Eq. 2.1). The energy required to create this area via bubble deformation is called the osmotic pressure Π_o. A thought experiment aiming at representing the osmotic pressure on a macroscopic level is shown in Fig. 2.4 [60].

This experiment consists in a liquid foam resting on a semi-permeable membrane letting the liquid phase through but not the bubbles. The semi-permeable membrane rests on a liquid reservoir and is free to move along the z-axis. If the liquid fraction of the foam is above the critical liquid fraction, i.e. $\varphi > \varphi_c$, the semi-permeable membrane moves spontaneously upwards and liquid is mechanically sucked out of the foam through the membrane, reducing the liquid fraction until the jamming transition is reached, i.e. $\varphi = \varphi_c$. One has then a liquid foam with close-packed spherical bubbles. In order to suck even more liquid out of the foam, which requires the deformation of the bubbles, one has to apply a force on the semi-permeable membrane equal to the osmotic pressure Π_o. The osmotic pressure can thus be expressed as

$$- \Pi_o \mathrm{d}V = \gamma \mathrm{d}A, \tag{2.10}$$

with $\mathrm{d}V$ being the volume of the liquid sucked out of the foam and $\mathrm{d}A$ the interfacial area created by the removal of liquid. The osmotic pressure Π_o varies with the foam height h. The liquid fraction profile is in equilibrium when the forces are in equilibrium, i.e. [60, 71]

$$\mathrm{d}\Pi_o = (1 - \varphi(z))\Delta\rho\, g\, \mathrm{d}z. \tag{2.11}$$

The osmotic pressure is obtained by integrating Eq. 2.11 over the liquid fraction profile $\varphi(z)$, which can be measured experimentally.

2.1.2 Stability of Liquid Foams

Liquid foams are thermodynamically unstable and age towards the lowest minimal interfacial area, i.e. a flat pool of liquid. Ageing occurs through four destabilisation mechanisms: evaporation, drainage, Ostwald ripening or coarsening, and coalescence.

Evaporation Often left out in the literature, evaporation plays, however, an important role in foam destabilisation. Evaporation can be limited by controlling humidity or by simply sealing the foam's container. Li et al. [70] provide a quantitative study on the influence of environmental humidity on the stability of aqueous foams and show how an uneven evaporation induces Marangoni instabilities which lead to bubble bursting.

Drainage Although we did not properly name this mechanism, we discussed it extensively in the previous part: drainage is the flow of liquid through the foam films, Plateau borders and vertices down the foam under the action of gravity. Drainage is what sets the liquid fraction profile and when it stops depends solely on the bubble size and the capillary length l_c. Indeed, increasing the viscosity of the liquid does slow down drainage but does not affect the equilibrium liquid fraction profile. The time required to reach this equilibrium, i.e. the time at which drainage stops, can be quantified as follows

$$\tau_d = \frac{\eta}{\varphi^\alpha R^2}, \tag{2.12}$$

with η being the viscosity of the liquid, R the average bubble radius and α a dimensionless exponent between 0.5 and 1 [19]. One can, however, hinder drainage by gelling or solidifying the liquid foam (see Sect. 2.3) or by blocking the Plateau borders using colloidal particles [58]. Note that drainage does not destroy bubbles: it changes their shape but neither their volume nor their number.

Ostwald ripening and coarsening Ostwald ripening is the direct consequence of the Laplace pressure in bubbles, and more specifically the differences in Laplace pressures between bubbles. The Laplace pressure p_L of a bubble in a liquid reads [28]

$$p_L = \frac{2\gamma}{R}, \tag{2.13}$$

with γ being the surface tension of the liquid and R the bubble radius. Bubbles strive towards a reduction of their Laplace pressure, and a straightforward way to do so is to increase the bubble radius. This is the driving force of Ostwald ripening, as the smaller bubbles—with a small radius and thus a high Laplace pressure—lose gas to the bigger bubbles until their dissolution. In other words, the small bubbles empty themselves into the big ones via transport of the gas through the continuous liquid phase, as illustrated in Fig. 2.5a. This results in a disappearance of the small bubbles and an increase of the average bubble size. One way to prevent Ostwald ripening is to use monodisperse foams: if all the bubbles have the same size, they also have the same Laplace pressure and there is no longer any pressure difference driving Ostwald ripening. One can also hinder Ostwald ripening by using particles at a high enough concentration so that they densely pack at the liquid-gas interface. The principle is to "seal" the liquid-gas interface with a densely packed particle layer and prevent the dissolution of gas into the liquid. One can also slow down Ostwald ripening by acting directly on the gas by choosing a gas which is poorly soluble in the liquid. For example, N_2 has a much lower solubility in water than CO_2, which means that N_2-containing foams will be more resistant to Ostwald ripening than CO_2-containing foams [121]. Likewise, one can add traces of an insoluble species, such as a fluorocarbon, in the gas phase. The presence of an insoluble species induces a chemical potential which is equal in all bubbles. The diffusion of gas through the liquid would modify this chemical potential, and one would lose the equality of the chemical potentials in the bubbles: the chemical potential acts against the Laplace pressure difference and prevents Ostwald ripening.

Coarsening lies on the same physics as Ostwald ripening and has the same consequences, but one typically uses the term "coarsening" for polyhedral foams, i.e. above the jamming transition, while "Ostwald ripening" is preferred for spherical bubbles in a liquid. Coarsening results from the gas exchange between the two sides of a foam film, as gas diffuses from the concave side of the film to the convex side of the film, as shown in Fig. 2.5b. One needs to sum the gas transfers over all the faces of the polyhedral bubble to know if it gains or loses gas. The gas transfer is dictated by the pressure difference $p_1 - p_2$ which results directly from the Laplace pressure $4\gamma/R$, R being the radius of curvature of the film [28]. As a result, there is no pressure difference between two sides of a perfectly flat foam film, and the highest pressure is on the concave side of the foam film. The characteristic coarsening time is inversely proportional to the permeability of the liquid to the gas and to the square of the bubble radius R^2 [94]. Moreover, it depends sensitively on the liquid fraction

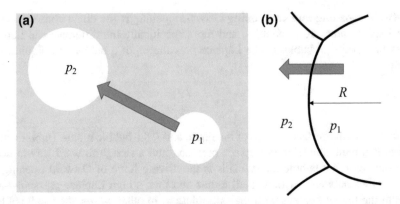

Fig. 2.5 Schematic representation of **a** Ostwald ripening between two bubbles in a liquid and **b** coarsening between two neighbouring bubbles in a low-density foam. The grey arrows show the direction of the gas flux from the smaller bubble to the bigger bubble and from the concave side of the foam film to the convex side of the foam film. R is the radius of curvature of the foam film (adapted from [19])

φ, since a higher liquid fraction implies a smaller surface area of thin films through which the gas can diffuse [61].

Coalescence Coalescence is the breaking of the thin film between two bubbles in contact and is directly related to the thin film stability. When a thin film bursts, the two bubbles in contact merge into a single bubble with a volume equal to the sum of the volumes of both bubbles. Coalescence is avoided by adding stabilising agents such as surfactants, polymers, particles, which adsorb at the gas-liquid interface [90].

2.1.3 Monodisperse Liquid Foams

Let us focus now on monodisperse foams, since most of the foams presented in the work at hand are monodisperse. A foam is monodisperse if its polydispersity index (*PDI*) defined as

$$PDI = 100 \cdot \frac{\sqrt{< d^2 > - < d >^2}}{< d >}, \tag{2.14}$$

with d being the bubble diameter, is below 5% [38]. Note that the same criterion applies to solid foams, the diameter d is then the pore diameter.

High-density foams Monodisperse foams tend to self-order under the action of gravity and confinement provided that their liquid fraction is high enough to allow for the bubbles to rearrange for a low energetic cost [38]. One usually observes two different crystalline structures in high-density liquid foams, namely FCC (face-centred cubic) and HCP (hexagonally close-packed) [116, 117]. Both packings consist in a

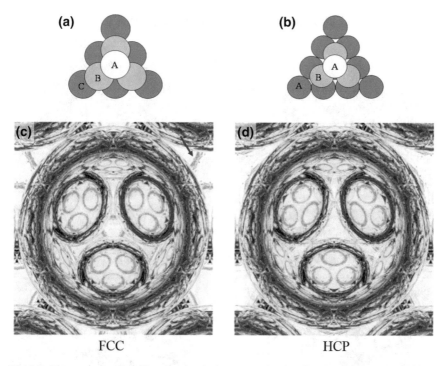

Fig. 2.6 The two main crystalline close-packed structures in monodisperse foams, FCC and HCP. **a** FCC packing structure over three layers with an ABC layering. **b** HCP packing structure over three layers with an ABA layering. Simulations showing the patterns in bubbles with **c** an FCC packing and **d** an HCP packing. The red rings outline bubbles from the layer below the top layer, and the red arrow shows the black line binding two bubbles in contact which is only observed in FCC packings (adapted from [117])

super-position of hexagonally close-packed monolayers, but while the FCC packing consists in a superposition of the monolayers in an ABC sequence, i.e. the third layers rests on a different position than the first layer, the HCP packing consists in a superposition of the monolayers with an ABA sequence, i.e. the third layer lies on the same position as the first layer, (see Fig. 2.6 a, b) [117].

One can easily distinguish both packings by looking closely at the liquid foams, as shown in Fig. 2.6c, d, provided that the foam is at least three-layers thick. Indeed, the simulated images show that one can distinguish three layers by looking at the top of the foam. The three rings observed in the top bubbles, one of which is circled in red, are present in both packings and do not differ. However, one sees three bubbles from the third layer within this ring. The way these rings are arranged tells us whether the foam has an FCC or HCP packing. If the third-layer bubbles are so that two bubbles are disposed outwards and one inwards, the packing is HCP, while if one bubble is disposed outwards and two inwards the packing is FCC. Another difference which can help confirm the packing observed is the presence of straight lines between two

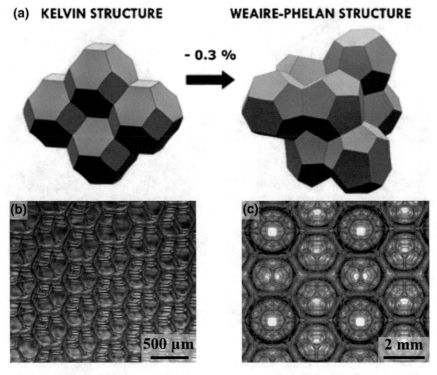

Fig. 2.7 a The energy gain from a Kelvin structure to a Weaire–Phelan structure [38]. **b** Experimental view of the Kelvin structure in a low-density liquid foam [117]. **c** Experimental view of the Weaire–Phelan structure in a low-density liquid foam [51]

neighbouring bubbles of the top layers, as pointed out by the red arrow in Fig. 2.6c. The straight lines form a triangle between three bubbles in contact which is only observed in an FCC packing. Note that the BCC (body-centred cubic) packing is not close-packed, therefore energetically unstable, which is why it is not experimentally observed in high-density foams [116].

Low-density foams For a long time, the most energetically favourable structure was thought to be the Kelvin structure, shown in Fig. 2.7a, c, which is a BCC packing in the low-density limit [5]. However, the Weaire–Phelan structure was proven to have a 0.3% lower surface energy than the Kelvin structure [120]. While all bubbles in the Kelvin structure have the same shape, the Weaire–Phelan structure has two different types of bubbles, namely 12-sided polyhedra and 14-sided polyhedra, as shown in Fig. 2.7.

Although the Weaire–Phelan structure has a lower energy than the Kelvin structure, it is very difficult to obtain experimentally. Indeed, repositioning the bubbles towards this structure requires an energy input that cannot be brought solely by thermal energy. The one example shown in Fig. 2.7c required the use of a mould

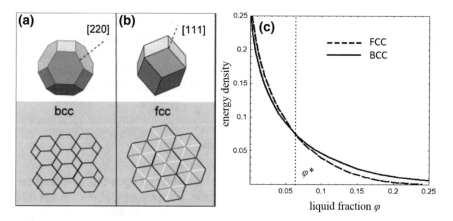

Fig. 2.8 **a** Kelvin cell or low-density BCC bubble with its projected structure on the plane normal to the [220] direction. **b** Rhombic dodecahedron or low-density FCC bubble, with its projected structure on the plane normal to the [111] direction. **c** Interfacial excess energy density (dimensionless) plotted as a function of the liquid fraction (adapted from [60])

imprinted with the Weaire–Phelan structure and strong shaking [51]. Therefore, the Kelvin structure remains the most often observed structure in low-density liquid foams, although, as mentioned earlier, it is not present in high-density liquid foams. The reason is that the low-density BCC structure (see Fig. 2.8a) results from the reorganisation of FCC-packed bubbles (see Fig. 2.8b) above the high-density limit through drainage [60].

The driving force for this structural change is the interfacial excess energy density, which above a critical liquid fraction $\varphi^* = 0.063$ is lower for an FCC packing but below φ^* is lower for a BCC packing (i.e. for a Kelvin structure). Figure 2.8c provides a thermodynamics-based explanation of why one does not usually observe high-density liquid foams with a BCC structure (i.e. high-density Kelvin cells), nor low-density liquid foams with an FCC structure (i.e. low-density rhombic dodecahedra). As a foam ages, it drains and its liquid fraction decreases. The bubbles reorganise from a high-density favoured FCC structure to a low-density favoured BCC structure. We have seen that the liquid fraction follows a gradient with foam height (see Sect. 2.1.1); van der Net et al. [117] showed the presence of FCC-packed bubbles at the bottom of a monodisperse liquid foam and BCC-packed bubbles at the top of the same liquid foam.

2.2 Polymer Foams

Polymer foams are analogous to liquid foams in that they are also dispersed systems, but the continuous phase is a solid polymer. As a result, polymer foams are thermodynamically stable, and one does not have to discuss their ageing—at least under atmospheric conditions and in the scope of the Thesis at hand. However, one needs to focus on the foam morphology, density and the relations that come at play between the structural properties and the mechanical properties. We will discuss in this chapter general principles on polymer foams and their mechanical properties, most of which also apply to other types of foams, e.g. metal foams or ceramic foams.[3]

2.2.1 Morphology of Polymer Foams

We have already seen that a polymer foam can be either open-cell or closed-cell, according to whether or not neighbouring pores are interconnected via holes (also called windows, openings or interconnects) in the pore wall (Fig. 1.1b, c). The pore connectivity is an essential parameter when discussing the morphological properties of a polymer foam. Indeed, pore connectivity influences greatly many properties of the polymer foam, e.g. the gas permeability or thermal conductivity. Moreover, the mechanical models one applies for polymer foams differ according to whether the foams are open- or closed-cell (see Sect. 2.2.3). One can measure the open-cell/closed-cell ratio by water absorption or permeation [43]. The openings form during the solidification of the liquid foams as the film between neighbouring bubbles ruptures (Sect. 2.3.3). The pore size distribution plays a huge role depending on the applications aimed for. For example, catalysis requires pores as small as possible to reach for the largest surface area possible, while tissue engineering requires pore sizes for the cells one wants to grow on the scaffold, i.e. pore sizes between 100 and 500 μm [76, 80]. Porous materials are classified according to the order of magnitude of their pore sizes. The classification proposed by the International Union of Pure and Applied Chemistry (IUPAC) is

- *micropores* are pores with diameters between 0.3 and 2 nm,
- *mesopores* are pores with diameters between 2 and 50 nm,
- *macropores* are pores with diameters bigger than 50 nm [106].

The shape of the pore is also relevant, especially in the case of anisotropic pores which induce anisotropic properties. Therefore, many polymer foams manufacturers characterise their products in the length and the width direction [43]. The thickness of the pore walls and the section of the struts also play an important role, especially regarding the mechanical properties of the polymer foam. In other words, what matters is how much of the material of the foam is in the struts and how much is in the films, as its contribution to the mechanical properties differs according to where the material is distributed (cf. Sect. 2.2.3).

---

[3]Much like in Sect. 2.1 on liquid foams and for the sake of brevity and clarity, we will not discuss the different processes existing for the generation of polymer foams. The reader who is interested in this topic will enjoy reading [76].

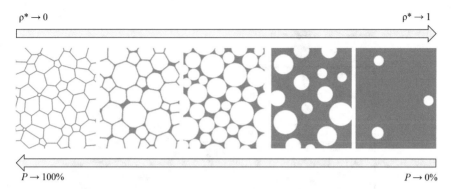

Fig. 2.9 Evolution of the relative density ρ^* and porosity P in a solid foam (adapted from [36])

2.2.2 Density of Polymer Foams

One has to be cautious when discussing the density of polymer foams, as one can define several densities for the same material. First of all, we can define the density of the polymer foam ρ_{foam} simply by the weight of the foam divided by its volume. But a polymer foam is composed of two phases, namely one polymer matrix with its own density ρ_{polymer} and a dispersed gas phase. The ratio of the foam and polymer densities is the relative density $\rho^* = \rho_{\text{foam}}/\rho_{\text{polymer}}$, which is equivalent to the liquid fraction φ of a liquid foam [54]. The relative density corresponds to the solid content in the foam and industrials amongst others prefer to talk about the gas content expressed in percentage by defining the porosity $P = 100 \cdot (1 - \rho^*)$. The relative density varies thus from 0 to 1, with $\rho^* = 1$ being the non-porous polymer, as shown in Fig. 2.9. Conversely, the porosity P varies from 0% for the bulk polymer to 100%. The relative density is of tremendous importance for the materials property, as the closer to 1 it gets, the more the foam's physical properties tend to the properties of the bulk polymer.

2.2.3 Mechanical Properties of Polymer Foams

The mechanical properties of polymer foams are most often characterised via compression tests such as illustrated in Fig. 2.10. One applies a force F to a polymer foam having a surface S, which induces a compression of the foam, which goes from an initial height l_0 to a height l with $\Delta L = l_0 - l$. During a compression test, one records the normal stress σ, defined as

$$\sigma = \frac{F}{S},\qquad(2.15)$$

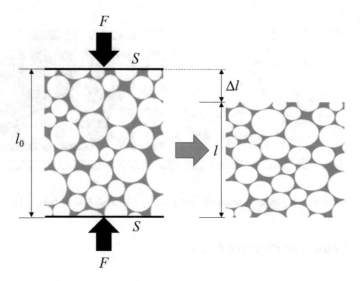

Fig. 2.10 Compression test on a polymer foam. A force F is applied on a foam with an initial height l_0 and a surface S. The sample shrinks under the stress applied to reach a height l

as a function of the strain ε, defined as

$$\varepsilon = \frac{\Delta l}{l_0}. \tag{2.16}$$

Experimentally, one can either set a stress ramp and record the strain at all stresses or set a strain ramp and record the stress at all strains.

One can plot a *stress-strain curve* from a compression test, which for porous polymers looks as sketched in Fig. 2.11. One can distinguish three main regions in which the polymer foams respond differently to compression.

In the linear elastic region, blue box in Fig. 2.11, the struts—and pore walls in the case of closed-cell foams—buckle elastically, i.e. they would recover their initial shape if one would remove the applied stress, and the whole foam would recover its initial volume. One speaks of a bending dominated behaviour [54]. The stress-strain curves follow a linear trend whose slope is the elastic modulus E. The elastic modulus constitutes a quantitative measurement for the stiffness of the foam, i.e. its resistance to deformation. The deformation is no longer elastic when it becomes irreversible [54]. The stress above which the foam is no longer in the elastic regime is the yield stress σ_y and it is close to the stress of the plateau region. The response of the foam in the plateau region, green box in Fig. 2.11, strongly depends on the nature of the polymer—whether it is elastomeric, plastic, brittle, or as often a combination of all three—and the morphology of the foam. Indeed, open-cell foams tend to have a bending dominated behaviour while closed-cell foams tend to have a stretch dominated behaviour [54]. The difference is caused by the relative amount of material

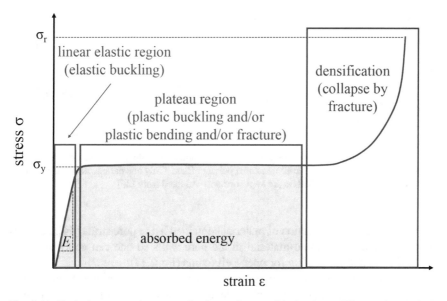

Fig. 2.11 Typical stress-strain curve of polymer foams with the three different characteristic regions, namely the linear elastic region, the plateau region and densification. The slope in the linear elastic region is the elasticmodulus E. The stress at which the linear elastic region ends is the yield stress σ_y and the stress at which the foam completely ruptures is denoted σ_r (adapted from [54])

in the struts and the pore walls. Indeed, the scaling equations for the description of the mechanical response of polymer foams in the linear elastic region differ for open-cell and closed-cell foams [54]. Scaling allows us to minimise the influence of the foam density and the elastic modulus of the polymer on the elastic modulus of the polymer foam in order to better observe the morphology-induced mechanical behaviour. On the one hand, for isotropic open-cell polymer foams, the scaling law reads

$$\frac{E_{\text{foam}}}{E_{\text{polymer}}} = \left(\frac{\rho_{\text{foam}}}{\rho_{\text{polymer}}} \right)^2 , \qquad (2.17)$$

which one can rewrite such that

$$E_{\text{foam}} \sim C \cdot \rho^{*2}, \qquad (2.18)$$

with C a constant. On the other hand, for closed-cell foams, one scales the elastic modulus according to

$$\frac{E_{\text{foam}}}{E_{\text{polymer}}} = \phi^2 \left(\frac{\rho_{\text{foam}}}{\rho_{\text{polymer}}} \right)^2 + (1 - \phi) \frac{\rho_{\text{foam}}}{\rho_{\text{polymer}}}, \qquad (2.19)$$

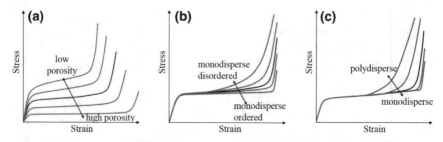

Fig. 2.12 Influence of **a** the porosity, **b** the degree of order and **c** the polydispersity on the shape of the stress-strain curve of a two-dimensional polymer foam. Only one parameter is varied in each sub-figure and all other parameters are kept constant. Adapted from [26]

where ϕ is the volume fraction of material contained in the pore struts [54]. For fully open-cell foam, there is no material in the pore walls so one can apply $\phi = 1$ to Eq. 2.19 to obtain the scaling for open-cell foams (Eq. 2.17).

In the plateau region, the foam can be submitted to elastic buckling and/or plastic bending and/or to a collapse by fracture. In the plateau region, the foam deforms irreversibly at an almost constant stress but keeps storing energy up until densification. The area below the stress-strain curve in the linear elastic and plateau regions is equal to the energy absorbed by the foam, i.e. the energy required to crack the material, which is a quantitative measurement for the toughness of the foam [54]. Densification results from the crushing of the foam as the different layers of pores in the foam collapse on top of each other and the density of the foam strongly increases. This results in a strong increase of the stress, and one can read a second elastic modulus which approaches the elastic modulus of the bulk polymer. The maximal stress or stress at rupture σ_r is a quantitative measure of the strength of the material, in other words, its resistance to failure [54].

As already mentioned, the shape of the stress-strain curve strongly depends on (a) the mechanical properties of the polymer constituting the foam, (b) the morphology of the solid foam and (c) the relative density of the polymer foam. Let us leave out the mechanical properties of the polymer constituting the foam and focus on the morphology and relative density of the polymer foams. Figure 2.12 displays the qualitative changes of the stress-strain curve of a polymer foam when modifying the foams morphological properties based on computer simulations of two-dimensional foams [26].

One sees in Fig. 2.12a that an increase of the relative density (or decreasing the porosity) results in an increase of the elastic modulus and the yield stress, i.e. the foam becomes stiffer. Increasing the relative density also leads to a shorter plateau which is also less flat, i.e. densification starts at a lower strain. The flatness of the plateau comes from the porous character of the foam and no plateau is observed in bulk polymers. In other words, when one increases the relative density, i.e. the amount of polymer in the foam, the mechanical properties of the foam tend to the mechanical properties of the polymer constituting the foam and its strength is improved.

Looking at morphological characteristics, one sees in Fig. 2.12b, c that neither the pore organisation, i.e. whether the pores are ordered or not, nor the polydispersity affects the linear elasticity of the polymer foam; only the relative density does. However, one sees that decreasing the degree of order of the pore or increasing the width of the pore size distribution induces an earlier densification and the foam can store less energy. At constant density, a monodisperse highly ordered foam should thus store the highest amount of energy and be the toughest possible. These results come from two-dimensional computer simulations but are interesting as it is a great challenge to experimentally verify these findings. Indeed splitting the different morphological parameters such as in Fig. 2.12 requires a perfect control over the foam morphology which has not yet been experimentally reached.

2.3 Foam Templating

Foam templating deals with the transition from liquid foams (cf. Sect. 2.1) to solid foams (cf. Sect. 2.2) and may help better understand the structure-properties relationships of solid foams. The following discussion is developed in more details in [4]. Some readers may draw the parallel between foam templating and emulsion templating, which leads to polyHIPEs (polymerized High Internal Phase Emulsion) [22, 86, 102, 103]. A polyHIPE is a porous polymer obtained from the solidification of the continuous phase of a high internal phase emulsion and the removal of the dispersed phase. Foam templating is similar to emulsion templating except that the dispersed phase is a gas. In emulsion templating, the emulsion can be either a water-in-oil or an oil-in-water emulsion, i.e. both polar monomers (oil-in-water emulsion) and apolar monomers (water-in-oil emulsion) can be used as a precursor for the targeted porous material. Moreover, the small density difference between the two phases leads to less ageing effects than in liquid foam templates. Work on polyHIPEs being very active over the last 10 years, a number of review articles have been written [64, 102–105], which may also serve as inspiration. The reason for the small number of studies published for liquid foam templating is due to the fact that this scientific field is still in its infancy.

2.3.1 Material of the Continuous Phase

One can identify three different types of liquid foam templates, namely monomer-based, polymer-based and dispersion-based foams, examples of which are shown in Fig. 2.13. In monomer-based systems, the solid foam is generated by polymerising monomers which are either present in the continuous phase of the liquid foam or constitute the continuous phase itself. Monomer-based liquid foam templates typically consist of an apolar monomer and an apolar gas, which makes the foams difficult to stabilise as surfactants are inefficient in such systems. Hence the resort to particle stabilisation (or Pickering stabilisation) [77] or to highly viscous solutions which

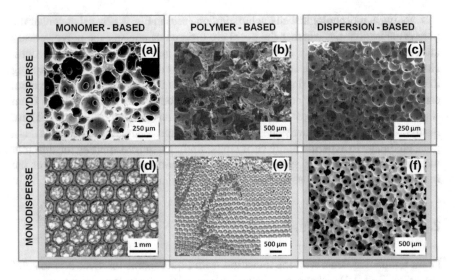

Fig. 2.13 Examples of solid polymer foams synthesised from liquid foam templates. **a** Polydisperse methylmethacrylate/dimethacrylate foam [88]. **b** Polydisperse chitosan foam [own unpublished work]. **c** Polydisperse polystyrene foam [97]. **d** Monodisperse acrylamide/N,N'-methylenebisacrylamide foam [118]. **e** Monodisperse chitosan foam [own unpublished work]. **f** Monodisperse styrene foam [87]. From [4]

make stable foams due to their high viscosity [67]. In polymer-based systems, solidification occurs by cross-linking a polymer melt [52] or a polymer dissolved in the continuous phase of the liquid foam [6, 7, 20, 23, 113]. Dispersion-based systems make use of a dispersion as continuous phase. This dispersion can either be a suspension (solid-in-liquid dispersion) or an emulsion (liquid-in-liquid dispersion). In the first case, solid polymer particles (suspension) are dispersed in a solvent which is removed for solidification, generally through drying. The foams are usually sintered to provide a sufficient cohesion between the particles. Suspension-based foams are either formulated so that they are stabilised by the same polymer particles (Pickering foams) [122], or they are stabilised by surfactants [84]. Well-known examples of suspension foams include natural and synthetic latex foams [16]. The second type of dispersion is an emulsion, i.e. it consists of droplets of an immiscible liquid in a continuous phase. Their foaming leads to foamed emulsions, or "foamulsions" which have been studied extensively over the last years [74, 95, 96]. The dispersed droplets contain the monomer whose polymerisation leads to the solid foam after removal of the solvent [45, 87].

2.3.2 Pore Size Distribution and Pore Organisation

Ideally, the pore size distribution of the solid foam is set by the bubble size distribution of the liquid template, i.e. the foam template is stable until solidification and the solidification process does not induce shrinkage or deformation. In such an ideal case, the pore size distribution and the bubble size distribution would be identical. However, this kind of perfect control is merely theoretical. In the laboratory, the challenge is to determine how much the pore size distribution deviates from the bubble size distribution in order to adapt the structure of the liquid template to the targeted solid foam. Indeed, one needs to account for foam ageing (Sect. 2.1.2). That is why one should, when possible, compare the morphology of the polymer foam with the morphology of the initial liquid foam. Bey et al. [13] showed that one can stabilise liquid foams against coarsening by combining the action of insoluble gas molecules in the bubbles and increasing the shear modulus of the foam via its gelation. They provide a complete phase diagram taking into account the influence of the different parameters on foam stability, i.e. the bubble size, the shear modulus of the foam and the insoluble gas content. When aiming at a specific pore size distribution, the choice of the foaming method is essential since a given foaming method gives access to a specific bubble size distribution [40]. Most of the commonly used foaming techniques for polymer foam templating provide little or even no control over the bubble size distribution, thus leading to polydisperse foams. Microfluidics provides the best control over the bubble size distribution (Sect. 2.4) and has therefore been adapted to complex fluids such as monomer solutions, polymer solutions or dispersions in order to produce well-controlled liquid foam templates for solidification [20, 23, 24, 45, 84, 87, 113, 114, 118]. A growing interest in monodisperse polymer foams has arisen in recent years, with a view to better controlling the morphology of solid foams. Being able to tailor the foam morphology also improves control over the foam's physical properties (e.g. mechanical, thermal) which can be set in accordance with the targeted applications.

2.3.3 Pore Connectivity

An interconnect comes about through the opening of the film separating two neighbouring bubbles. As discussed in Sect. 2.1.2, the mechanisms controlling film rupture in foams have been extensively investigated in liquid foams. While certain trends have been evidenced, a coherent picture of film rupture in liquid foams remains yet to be developed. Systematic investigations in liquid template foams are few and far between. Figure 2.14 shows a series of polymer foams with open-cell and closed-cell foams made via foam templating.

The few examples of fully closed-cell materials existing in the literature [77, 122] are polymer foams which are made from liquid foams stabilised by particles. Particle-stabilised foams are peculiar in that they do not require any surfactant as the foam film is stabilised by adsorption of particles at the gas-liquid interface. Once adsorbed at the interface, it requires a high amount of energy to desorb the particles [15]. Such a strong anchoring of the particles at the gas-liquid interface renders the foam film more difficult to break and even to deform, depending on how densely the particles are packed. As a result, it is not surprising that particle-stabilised foams

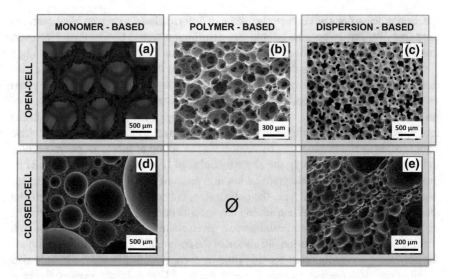

Fig. 2.14 Examples of solid polymer foams synthesised from liquid foam templates showing open-cell and closed-cell structures. **a** Monodisperse poly(acrylic acid) foam [112]. **b** Hyaluronic acid scaffold [7]. **c** Polystyrene foam from a foamed styrene-in-water emulsion [45]. **d** Macroporous epoxy resin [67]. **e** Pentafluorostyrene-Divynilbenzene foam [77]. No example of a closed-cell polymer-based foam was found in the literature (taken from [4])

do not yield polymer foams with open pores. Conversely, there is yet, to the best of our knowledge, no example of closed-cell polymer-based foams. Polymer-based foam templates preferably yield open-cell polymer foams. The main explanation is that the dispersed polymer tends to be expelled out of the film during its thinning, leaving a thin film without any material to solidify. Moreover, a standard method for studying the pore morphology in polymer foams is scanning electron microscopy (SEM), which requires that the samples are under a high vacuum. This vacuum may induce a rupture of thin films within the material, resulting in the observation of openings between pores when the foam is in fact closed-cell [37].

Interestingly, van der Net et al. [118] were able to obtain closed-cell foams with a system yielding primarily open-cell foams by accelerating the polymerisation. The authors do not provide any further explanation, but this shows that film rupture competes with polymerisation. The most intensive work on the question of cell-opening in liquid foam templates has been conducted by Zhang, Macosko and co-workers [125–127] on polyurethane foams. They used the non-solidifying templates without isocyanate as model system and showed the importance of the viscoelastic properties of the polymer-gas interfaces. Moreover, they revealed the decisive role of urea-formation during formation in conjunction with internal stresses to promote cell-opening during solidification. Another study by Testouri et al. [114] investigated the transition between fully open-cell foams and fully closed-cell polyurethane foams which were physically foamed. Their preliminary investigations show the intermediate stages between the two limits of pore connectivity. The authors show how increasing the liquid fraction/density and decreasing the solidification time leads to closed pores. Globally one obtains open-cell foams by decreasing the liquid fraction/density and increasing the solidification time. The authors hypothesise that during foam ageing the liquid is sucked out of the film which gets thinner until it breaks. Starting with a higher liquid fraction leads to a foam with thicker films and delays the point of film rupture. If the foam solidifies before this point, the solid foam obtained is closed-cell. If the foam solidifies after film rupture, the solid foam obtained is open-cell. The fact that different degrees of pore connectivity can be obtained suggests that the rupture of the film is slow and may occur during solidification. Poulard et al. [84] studied the size of the holes on polymer monodisperse foam monolayers made from aqueous latex dispersions stabilised by a surfactant. Interestingly, the size of the holes does not depend on the latex concentration in the foam, but varies linearly with the bubble size. The authors explain this dependency with the fact that the latex particles are expelled from the foam films since solidification is slow and the particles larger than the equilibrium thickness of the aqueous foam film. The film is, therefore, a purely aqueous film stabilised by surfactant which cannot solidify and breaks at a later stage of drying. One may therefore distinguish between two main classes of systems. When working with dispersion-based systems or polymer solutions, the dispersed objects/polymers may leave the thin film under certain conditions, leaving behind a pure solvent film which cannot be solidified. In this case, the pore opening corresponds to the thin film of the liquid template. In the second class of systems, the separation of a solidifying element from a solvent is not possible and the film rupture, therefore, becomes a question of matching timescales of film thinning and solidification [112].

2.3.4 Porosity and Material Distribution

When polymer foams are made from a liquid foam, their porosity corresponds to the gas fraction $(1 - \varphi)$ of the liquid foam. Therefore, if one wants to tailor the porosity of a polymer foam, one needs to tailor the liquid fraction of the foam template. The porosities of foam templated polymer foams typically lie within the wide range of 50–90%, but unfortunately many studies do not compare the liquid fraction of the liquid foam template with the porosity of the solid foam [4]. One may think first of microfluidics as a way of tuning the liquid fraction, which allows for the fine-tuning of the liquid and gas flow rates, i.e. the liquid fraction. However, liquid foams are usually subject to drainage and the liquid fraction follows a gradient (Sect. 2.1.1). One can calculate below which foam height h drainage does not occur and thus approximate the liquid fraction to be constant below this height h (Eq. 2.9). Looking at Eq. 2.9, one sees that the lower the bubble size, the higher is h, i.e. the larger the bubbles, the more important is drainage.

Figure 2.15 shows how important drainage can be in a monodisperse solid foam generated via microfluidics, here a chitosan cross-linked foam. As one can see the drained phase is as high as the foam. The zoom-in shows the monodisperse hydrogel

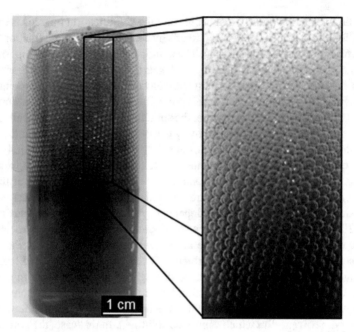

Fig. 2.15 Picture of a monodisperse genipin cross-linked chitosan foam showing the continuous phase issued from drainage. The zoom shows the cross-linked foam with a density gradient which is close to the liquid fraction gradient observed in monodisperse liquid foams (taken from [4])

foam with spherical bubbles at the bottom and polyhedral bubbles at the top of the foam. This structure is typical of monodisperse liquid foams and obeys the liquid fraction gradient imposed by foam drainage (Sect. 2.1.1). Although one can easily tailor the relative amount of gas and liquid injected with the help of microfluidics, this only determines the maximum liquid fraction which becomes irrelevant as soon as the foam starts draining. However, one can think of two objections to this assessment. (1) If the amount of liquid is increased in the initial liquid foam, the amount of liquid that drains is larger and thus it takes a longer time until the foam reaches its equilibrium liquid fraction. This is not relevant if allows all forces to equilibrate, but plays a role if the foam solidifies before this equilibrium is reached. This brings us back to the main idea of foam templating, i.e. the adjustment of the characteristic times inherent to foam ageing and foam solidification. (2) By setting the initial liquid fraction such that it is lower than the equilibrium liquid fraction, the foam does not drain and one can fine-tune the liquid fraction with microfluidics. This holds true, however, only for foams which are below the equilibrium liquid fraction. However, a side effect of working with low-density foams is that the driving forces leading to the ordering of the bubbles in monodisperse liquid foams do not overcome the interfacial energy of the foam, and one obtains a foam that is, although still monodisperse, disordered.

2.4 Microfluidics for Foam Generation

Microfluidics is the manipulation of fluids within micrometric dimensions. Microfluidics originates from the miniaturisation of many processes such as chemical and biomedical analysis. The name Lab-on-a-Chip is used to describe the chips in which micrometric channels allow for fluid mixing, fluid sorting, and analysis, amongst others [69]. One thus needs to understand fluid mechanics at the micrometric scale to understand the behaviour of flows in microfluidic channels. The small distances at play in microfluidics make the fluid mechanics involved peculiar in the way that one may discard the action of gravitation and inertia [111]. Physicists often use dimensionless numbers to characterise flows at the microscale, the most important one being the Reynolds number Re

$$Re = \frac{\rho v D}{\eta} = \frac{\text{inertial stress}}{\text{viscous stress}}, \qquad (2.20)$$

where ρ is the fluid viscosity, η the fluid viscosity, v the velocity of the fluid and D a characteristic dimension of the system [66]. The Reynolds number is relevant for it allows us to sort out the type of flow dealt with. For high Reynolds numbers, i.e. $Re > 10^5$, the flow is turbulent. The flow is not stable, and the velocity at a given point in space is not only not constant, but it also cannot be analytically predicted. Air flows in the atmosphere are turbulent, with $Re \approx 10^{11}$ [92], as well as smoke formed by a burning candle, which becomes turbulent from a given height, as seen in Fig. 2.16a. For low Reynolds numbers, i.e. $Re < 2000$, the flow is laminar and viscosity-driven as inertia can be safely neglected. Microfluidic flows are laminar. Practically, this means that the mixing of two phases within a microfluidic channel is not mechanical and can only occur through diffusion, as shown in Fig. 2.16b.

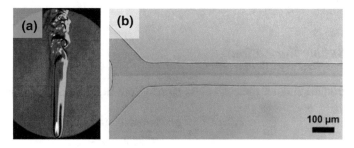

Fig. 2.16 Visualisation of flows at different Reynolds numbers. **a** Candle plume showing a transition from a laminar flow to a turbulent flow as the smoke gets further from the candle. The characteristic length D increases and Re increases with it until the laminar-turbulent transition occurs [99]. **b** Example of a co-flow in a microfluidic channel with two dye solutions flowing along the channel but not mixing due to their low Re [81]

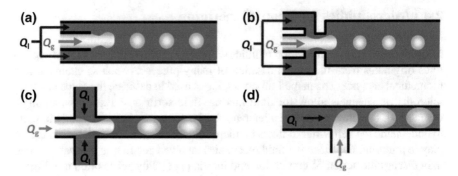

Fig. 2.17 Various examples of bubbling geometries: **a** co-flow geometry, **b** flow-focusing geometry, **c** cross-flow geometry and **d** T-junction. Note that flow-focusing is a specific case of co-flow with a constriction. The T-junction is also a particular case of cross-flow geometry. Q_l is the liquid flow rate and Q_g is the gas flow rate. [40]

2.4.1 Bubbling Geometries

Due to the stability of the flows involved, which have a low *Re*, microfluidics is the tool of choice for the production of monodisperse bubbles. Microfluidic bubbling allows for a periodic breakage of the gas flow by the liquid flow, yielding same-size bubbles. But as seen in Fig. 2.16b, two fluids may flow along without bubble or droplet formation. The rupture of the dispersed phase thus requires in most cases a geometry-induced constraint. Several geometries may be used for microfluidic bubbling: co-flow, flow-focusing, cross-flow and T-junction (Fig. 2.17).[4] The geometry is important in that it sets the flow field near the point where the two phases meet and thus fixes the dynamic forces involved in the gas break-up leading to bubble formation [40].

2.4.2 Bubble Formation

Let us introduce another dimensionless number, the Bond number *Bo*, defined as [10]

$$Bo = \frac{\Delta \rho\, g\, D^2}{\gamma} = \frac{\text{gravitational stress}}{\text{interfacial stress}}, \qquad (2.21)$$

[4]This section deals with the generation of bubbles via microfluidics. The reader should however know that replacing the gas phase with a liquid of opposite polarity to that of the continuous phase leads to the generation of droplets, and thus monodisperse emulsions. We will discuss here only examples of monodisperse bubbling, but bear in mind that droplet-based microfluidics rests on the same physics and the present discussion about bubbles holds true for droplets.

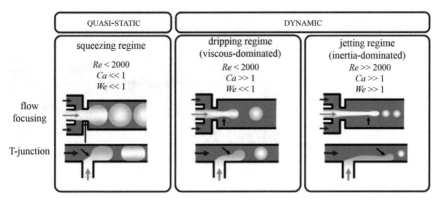

Fig. 2.18 The three bubbling regimes accessible with a flow-focusing geometry and a T-junction. The grey arrows show the directions of the gas and liquid flows and the black arrows show the regions where bubble break-off occurs (adapted from [40])

where $\Delta\rho$ is the density difference between the continuous and the dispersed phase, g is the gravitational acceleration and γ the interfacial tension between both phases, i.e. the surface tension of the liquid if one works with a liquid and a gas phase. In microfluidics, the fluids are confined within channels of micrometric dimensions and thus $Bo \ll 1$, i.e. one can neglect the influence of gravity in microfluidic processes. One differentiates three bubbling regimes depending on the relative importance of the different stresses involved, namely viscous stresses, interfacial stresses and inertial stresses. The relative importance of the different stresses is quantified by dimensionless numbers. The capillary number Ca is used to compare interfacial and viscous stresses and is defined as

$$Ca = \frac{\eta\, v}{\gamma} = \frac{\text{viscous stress}}{\text{interfacial stress}}. \tag{2.22}$$

The Weber number We is used to compare inertial and interfacial stress and is defined as

$$We = \frac{\rho\, v^2\, D}{\gamma} = \frac{\text{inertial stress}}{\text{interfacial stress}}. \tag{2.23}$$

The three main bubbling regimes are shown in Fig. 2.18 through the scope of the flow-focusing geometry and the T-junction, which are the geometries used in the thesis at hand.

At constant geometry, the bubbling regime depends on the flow parameters, i.e. the gas and liquid flow rates and their ratio, which is contained the relevant dimensionless numbers Re, Ca, and We.

Squeezing regimef height h The squeezing regime is observed for low liquid flow rates, i.e. low Re, low Ca, and low We. The squeezing regime is quasi-static in that viscous stresses and inertial stresses do not play any role and the bubble size

Fig. 2.19 Bubble formation with a flow focusing geometry in the squeezing regime. The scale bar is 500 μm

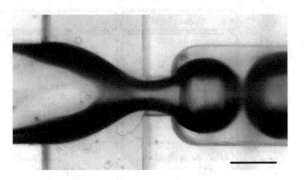

depends solely on the ratio of the liquid and gas flow rates [53]. The bubble is formed as the gas blocks the channel, and the liquid phase accumulates in the constriction for flow-focusing geometries, as it is seen in Fig. 2.19, or in the main channel for T-junctions until the built-up pressure pinches-off the bubble neck. At constant flow rates, this pinch-off is periodic and the bubbles produced have all the same sizes and one obtains monodisperse foams.

The bubble volume V_b can be predicted by the following law

$$V_b = V_o \left(1 + \alpha \frac{Q_g}{Q_l} \right), \tag{2.24}$$

where V_o is the critical volume from which the gas thread starts to block the section of the orifice/channel, and α is a dimensionless constant related to the geometry of the system [38]. From Eq. 2.24 one can easily determine the bubbling frequency f_b, i.e. the number of bubbles produced per second, namely

$$f_b = \frac{Q_g}{V_b}. \tag{2.25}$$

Since the bubble volume V_b strongly depends on the geometry the squeezing regime yields bubbles which are at least as large as the size of the constriction/channel, i.e. from a few hundreds of micrometres to several millimetres. The squeezing regime allows for the formation of very low-density foams with a low liquid fraction φ. However, the production rate of such foams is very slow. Therefore, the squeezing regime is not optimal for the production of monodisperse foams, despite its ability to produce foams with the lowest polydispersity [53] as seen in Fig. 2.21.

Dripping regime If one increases the flow rates, one increases the fluid velocity and therefore Ca and We to reach the dripping regime (Fig. 2.18). In the dripping regime, the viscous stresses overcome the interfacial stresses ($Ca \gg 1$) and the gas thread break-up without having to obstruct the orifice/channel, as shown in Fig. 2.20. The contribution of the viscous stresses is obvious when looking at Fig. 2.20, which shows that the gas thread is the thinnest at the entrance of the orifice, i.e. where the viscous stresses are the largest.

Fig. 2.20 Bubble formation with a flow focusing geometry in the dripping regime. R_o is the characteristic length of the orifice and R the bubble radius. The scale bar is 500 μm

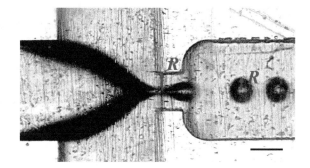

The bubble size R can be scaled with the characteristic dimension of the orifice R_o and the capillary number Ca of the liquid phase

$$R \sim \frac{R_o}{Ca}. \tag{2.26}$$

The bubble size becomes, as opposed to the squeezing regime, viscosity-dependent, which makes the system more sensitive to instabilities. Moreover, as seen in Fig. 2.21, the bubble size dependency on the gas flow rate is stronger than in the squeezing regime and small pressure variations induce a larger change in bubble size. Therefore, the monodispersity in the dripping regime is lower than in the squeezing regime. The dripping regime is, however, prefered for the generation of monodisperse foams as its production rate is higher than in the squeezing regime and it gives access to lower bubble sizes, i.e. below the size of the orifice, as one can see in Fig. 2.20.

Jetting regime The jetting regime is reached upon a further increase of the flow rates until inertia overcomes interfacial and viscous stresses (Fig. 2.18). The formation of bubbles via the jetting regime originates from the Rayleigh–Plateau instability, omnipresent in our daily-life under the form of an ink thread breaking into droplets on paper or droplets forming at the end of a water thread from a faucet [28]. Strong inertial stresses induce the formation of a long cylinder of the continuous phase, which breaks up into several bubbles in order to reduce its surface, and thus its surface energy (Sect. 2.1.1) [28]. The jetting regime is, however, not wanted for the generation of mono-disperse foams as the gas thread leading to bubble generation is highly sensitive to any small pressure variation and it yields foams with large polydispersities compared to the dripping regime or the squeezing regime (Fig. 2.21).

2.5 Chitosan and Its Hydrogels

2.5.1 The Chitosan Molecule

Chitosan is a polysaccharide derived from chitin, which is extracted from the shells of crustaceans [78]. The chitin and chitosan molecules are shown in Fig. 2.22. Chitosan

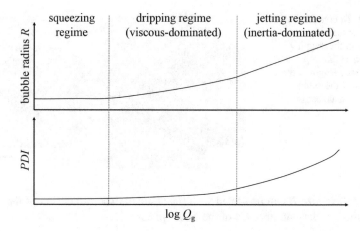

Fig. 2.21 Variation of the bubble radius R and polydispersity index PDI as a function of the logarithm of the gas flow rate $\log Q_g$ from the squeezing regime to the dripping regime and the jetting regime (adapted from [40])

Fig. 2.22 a Chemical structure of the chitin molecule and **b** the chitosan molecule showing acetylated and deacetylated units

has already been studied for applications such as tissue engineering and drug release [32, 41]. Indeed, thanks to its amino side groups, chitosan is intrinsically antibacterial and displays little reaction with foreign bodies [72]. The chitosan macro-molecule is composed of acetylated units and deacetylated units with an amino group. Chitosan is produced via the deacetylation of chitin. One speaks of chitosan once at least 60% of the units are deacetylated [78]. The deacetylation degree DD characterises a given chitosan macromolecule and is defined as the percentage of amino groups over all units of the macromolecules, i.e. chitosan has a DD of at least 60%. Chitosan is slightly soluble in acidic solutions, in which its amino groups are protonated (its amino group has a pK_a of 6.3) [12]. Chitosan is thus a polycation in acidic conditions and displays the common properties of polyelectrolytes [9]. Its charge density depends on the deacetylation degree since it corresponds to the number amino groups available [89].

Surface activity of chitosan The polycationic character of chitosan is of tremendous importance for the behaviour of the polymer in solution. Indeed, the charge density sets the strength of the electrostatic repulsion—at constant salt concentration—within the molecule itself, and, therefore, also sets the conformation and the stiffness of the

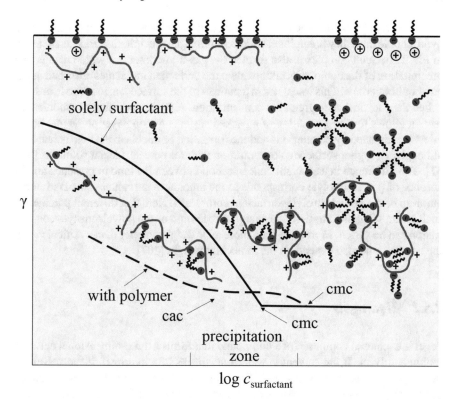

Fig. 2.23 Liquid-gas interface in the presence of a polycation and of an anionic surfactant. The graphic shows the variation of the surface tension γ with the surfactant concentration $c_{\text{surfactant}}$. The increase of the surfactant concentration is also schematised in the sketch from left to right. The full line represents the evolution of the surface tension if solely anionic surfactants were present in solution, while the dashed line represents the evolution of the surface tension of a solution containing an ionic surfactant and a polycation (redrawn from [55])

chain in solution [34]. The scaling theory of polymers, and more specifically poly-electrolytes, in solution will not be addressed here, but we recommend the reading of [27, 34, 35, 93] to the interested reader. We will focus on the surface activity of a polyelectrolyte in the presence of an oppositely charged surfactant. Goddard et al. [55–57] brought a large understanding on that topic. Let us develop the example of the surface activity of a polycation such as chitosan in the presence of an anionic surfactant. The polymer-surfactant interactions can be monitored via surface tension measurements and the determination of the cmc as shown in Fig. 2.23.

At low surfactant concentrations, some surfactant molecules are adsorbed at the gas-liquid interface, with some polyelectrolyte molecules being also adsorbed due to electrostatic interactions. The resulting polyelectrolyte-surfactant complex may be more surface active than the pure surfactant at equal concentration, which result in a lower surface tension in presence of polycation, as seen in Fig. 2.23. Upon addition of surfactant, the polyelectrolyte-surfactant complexes become more and more hydrophobic until they precipitate, which occurs in the precipitation zone as sketched in Fig. 2.23. Such complexes also form at the gas-liquid interface which results in a precipitation of the complexes taking down the surfactant molecules adsorbed at the gas-liquid interface. This translates in a plateau of the surface tension whose onset is called the cac, critical aggregation concentration. A further addition of anionic surfactant allows for the generation of a packed surfactant monolayer at the gas-liquid interface screened by counterions and the formation of micelles: the cmc is reached, although at a higher surfactant concentration than for pure surfactant solutions [55–57]. The difference in surfactant concentration between the cmc in presence and in absence of polyelectrolyte corresponds to the amount of surfactants involved in the build-up of the polyelectrolyte-surfactant complexes. Note that different parameters such as the salt concentration, the charge distribution or the polyelectrolyte concentration can be varied. Moreover, for a molecular weight smaller than a critical value, the cac and cmc also depend on the molecular weight [107].

2.5.2   Hydrogels

A gel is a material composed of a compound that forms a three-dimensional network within a solvent. If the solvent is water one speaks of a hydrogel, if the solvent is organic one speaks of an organogel. The compound constituting the network can be polymers [3], fibres or particles which organise themselves into a network, such as nanoclay [31]. We will, however, focus here on polymer-based systems since these are dealt with in the Thesis at hand. One may distinguish two types of gels, namely physical gels and chemical gels. In a physical gel the nodes between the polymer molecules, named cross-links, are physical and reversible, e.g. via heating [62]. The cross-links in chemical gels are built by a cross-linker which can react on at least two sites with a chemical function carried by the polymer. At the sol-gel transition in a chemical gel is irreversible due to the covalent bonds involved in the cross-linking process [62].

The process of gelation is well described by the percolation theory developed by de Gennes [27]. Let us consider the polymer solution to be cross-linked as an array of points, each point being a chemical function on a polymer chain being able to react with a cross-linker molecule.[5] As the cross-linking reaction starts, neighbouring polymer chains will be bonded via cross-linkers, as sketched in Fig. 2.24, left. As cross-linking goes forth, clusters build up and grow, as sketched in the centre of Fig. 2.24, up until enough cross-links are formed to constitute a large cluster spanning the entire system (Fig. 2.24 right): the percolation point is reached and a gel is formed.

---

[5]We develop the percolation theory here in the frame of a chemical gelation such as what is dealt with in the thesis at hand. Physical cross-linking is based, however, on the same principles and percolation can be described the same way for physical gels.

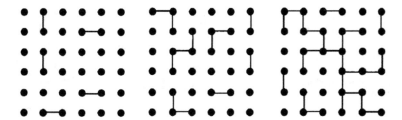

Fig. 2.24 Modelisation of the percolation theory by an array of points able to cross-link. The cross-linking degree increases from left to right until percolation is reached (adapted from [62])

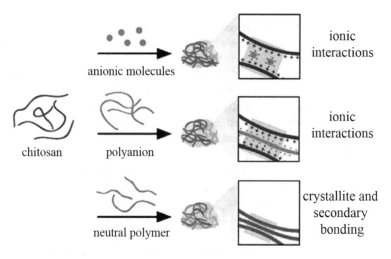

Fig. 2.25 Schematic representation of physical hydrogels based on chitosan with the different types of interactions leading to physical cross-links, depending on the second component added to induce gelation (adapted from [14])

Note that reaching percolation does not imply an arrest of the cross-linking reaction and more cross-links may form and strengthen the network [62].

Chitosan hydrogels Chitosan can form physical hydrogels via the addition of a second component such as anionic molecules, polyanions, or neutral polymers, as shown in Fig. 2.25.

Small ionic molecules such as sulphates, phosphates [100, 101] or metallic ions such as Pt(II) or Pd(II), which cross-link more via coordination bonds than electrostatic interaction [17], can induce physical gelation. The strength of the physical network, i.e. the density of cross-links, depends on the charge density and the size of the anionic compound as well as on the number of available protonated amino groups [14]. Physical gelation with addition of polyanions is based on the same ionic interactions. Polyanions such as DNA, polysaccharides (e.g. alginate), or proteins (e.g. keratin, collagen), have been used to build chitosan-based networks. Chitosan-polyanion networks are reversible and their mechanical strength depends on the

Fig. 2.26 **a** Chemical structure of genipin and **b** chemical structure of a chemical cross-link in a genipin-cross-linked chitosan hydrogel (redrawn from [79])

charge density, ionic strength, pH, and temperature [115]. Polymer blends of chitosan and another neutral polymer may also lead to physical cross-links via the formation of crystallites. PVA is a well-known example of neutral polymer being able to crystallise locally within a chitosan matrix [14]. Ladet et al. [65] showed that chitosan can self-cross-link upon addition of sodium hydroxide, which neutralises the amino groups of chitosan and hinders electrostatic repulsion within the chitosan chain. This promotes hydrogen bridges as well as hydrophobic interactions and allows for the chitosan chains to locally crystallise, inducing physical gelation.

Chitosan's amino side groups also provide numerous possibilities for making chemical hydrogels, as both the amino and the hydroxyl groups constitute a reactive site for cross-linking [8, 9, 25]. Known cross-linkers for chitosan are small molecules (e.g. glutaraldehyde, genipin), photo-sensitive molecules (e.g. functional azides), enzymes (e.g. phloretin hydrolase which produces phloretic acid) or polymers carrying reactive groups [14].

Let us focus on the specific case of chitosan cross-linked with genipin. Genipin is a biobased compound extracted from the gardenia plant [33] whose chemical structure is shown in Fig. 2.26a. The structure of a cross-link in a genipin-chitosan hydrogel is shown in Fig. 2.26b. Several mechanisms for the cross-linking reaction have been proposed. However, a consensus is lacking [18, 63, 75], but it has been shown that the cross-linking reaction is highly pH-dependent [29]. The cross-linking reaction induces a blue colour which one attributes to the oxygen radical-induced homopolymerisation of genipin and the reaction of this oligomer with amino groups [18]. As a result two chitosan molecules can also be cross-linked by a genipin oligomer and not only a single genipin molecule as described in Fig. 2.26b [79].

The blue colour is due to the presence of oxygen in air and is thus more intensive near the surface of the hydrogel. However, it allows for a macroscopic and simple monitoring of the cross-linking reaction [68]. Note that genipin reacts with unprotonated amino groups, while a good dissolution of chitosan in an aqueous solution requires the protonation of the amino group. Therefore, one has to find a compromise between a pH low enough to allow a good dissolution of chitosan and a pH high enough to provide enough reacting sites for the cross-linking reaction with genipin.

2.6 Cellulose Nanofibres and Nanocomposites

2.6.1 Cellulose Nanofibres

Cellulose is the most abundant polysaccharide in nature as it constitutes the structural skeleton of most of the vegetal biomass [85]. A mesh of cellulose nanofibres (CNFs), or microfibrils,[6] with a diameter of about 10–50 nm embedded in a hemicellulose and lignin matrix makes up the cell wall of green plants, as seen in Fig. 2.27. The cellulose microfibrils have hydroxide groups on their surface and are composed of cellulose chains arranged around the same axis which strongly promotes hydrogen bonding and results in a high degree of crystallinity [47, 59, 82]. The semicrystalline morphology of the cellulose chains provides rigidity to the microfibrils which thus possess an elastic modulus approaching the one of a perfect cellulose crystal. The presence of crystallites renders the cell wall rigid and not easily stretchable [47].

CNFs can be extracted from numerous plants via a first alkaline hydrolysis or bleaching step followed by a mechanical treatment such as high-intensity ultrason-ication, high-pressure homogenisation or grinding, amongst other processes [91]. The obtained microfibrils are a train of cellulose crystallites linked by amorphous cellulose regions. The crystalline regions provide mechanical strength to the fibres while the amorphous regions provide flexibility. The amorphous regions can, how-ever, be degraded via acid hydrolysis followed by centrifugation, sonication, and dialysis, yielding isolated crystallites called cellulose nanocrystals (CNCs) or cel-lulose nanowhiskers. While CNFs have diameters ranging from 10 to 100 nm and lengths of several micrometres (depending on the plant it was extracted from), CNCs have a low aspect ratio, a diameter ranging from 4 to 25 nm, and a length between 100 and 500 nm [91].

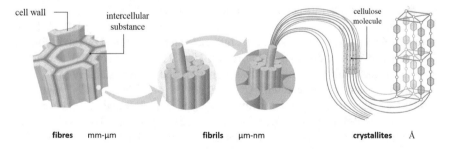

Fig. 2.27 Different degrees of association of cellulose in the cell wall of plants, starting from a mesoscopic plant fibre. The secondary wall is made of cellulose fibrils dispersed in an hemicellu-lose/lignin matrix. These fibrils consist of semicrystalline cellulose chains, i.e. with crystallites and amorphous regions, parallel to their axes (adapted from [128])

---

[6]The denomination "microfibrils" is more often used by the wood chemists community. When used as fillers for composite materials, however, the composite scientists prefer the term "cellulose nanofibres".

CNF can be modified for specific applications since the high density of hydroxide groups on the surface of the fibrils renders the fibrils highly negatively charged, which can be unwanted for some applications [83]. Quarternised cellulose nanofibrils have thus been developed to broaden the range of applications of CNFs. Quarternised CNFs, in which the hydroxide groups have been replaced by quaternary ammonium groups, can be prepared via a treatment of the fibres with glycidyl trimethylammonium chloride in the presence of NaOH [83].

2.6.2 Cellulose Nanofibre-Based Nanocomposites

Nanocomposites are materials composed of two components: a matrix and a filler, the filler having at least one dimension in the nanometre range. The matrix can either be a polymer, a ceramic material or a metal, and the filler may also be metallic, polymeric or mineral [2]. In the Thesis at hand, we are interested in polymer-based and polymer-filled nanocomposites. In other words, a polymeric filler is dispersed in a polymer matrix. Let us discuss such materials using the example of chitosan-based materials reinforced with CNF. Yano et al. [124] established in 1997 the potential of cellulose nanofibres to strengthen wood-based materials, opening the way to CNF-based nanocomposites [11, 44, 59, 98, 110, 123].

The main challenge in a composite material is to bind the filler to the polymer matrix to obtain the best reinforcement possible. Indeed a bad adhesion of the filler to the polymer matrix or its uneven dispersion within the matrix may not lead to the improved mechanical properties sought for [2]. Conveniently, chitosan and CNF can be both dispersed in aqueous solutions, which facilitates the interpenetration of the chitosan chains and the CNF. As shown in Fig. 2.28, Wang et al. [119] developed a templating method towards chitosan-CNF nanocomposite foams by freezing and freeze-drying a mixed dispersion of chitosan and CNF. This method is known as ice templating, and the pores result from the sublimation of ice crystals [21]. The pore morphology and anisotropy of the material can be fine-tuned by modifying the freezing conditions of the liquid mixture [73, 109]. The addition of CNF to chitosan yields materials with better mechanical properties and thermal stability [119].

Chitosan foams reinforced with CNF are not the only chitosan-CNF nanocomposites that one can produce. Fernandes et al. [49, 50] generated transparent CNF-filled chitosan films by casting mixed chitosan-CNF suspensions. The resulting films show improved mechanical properties and thermal properties compared to the CNF-free system. Note that the dispersion for the generation of such nanocomposite films is similar to the one described above (Fig. 2.28). Therefore, one can imagine as many CNF-filled chitosan nanocomposite materials as there are chitosan-based materials. Many examples of chitosan-CNF nanocomposites, and more generally polysaccharide-CNF nanocomposites, are described in the following reviews [1, 11, 50].

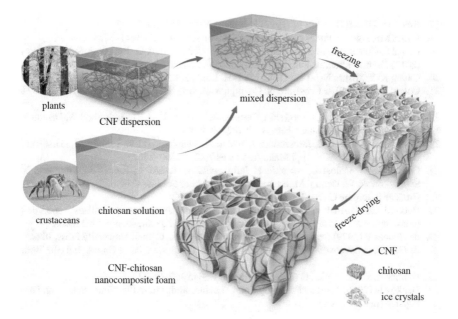

Fig. 2.28 Procedure for the generation of chitosan-CNF nanocomposite foams via ice-templating. An aqueous CNF dispersion is mixed with a chitosan solution. The mixture is subsequently frozen to yield a solid in which chitosan and CNF are expelled out of the ice crystals. The ice crystals sublimate during freeze-drying to yield a porous chitosan-CNF nanocomposite [119]

References

1. Abdul Khalil H, Bhat A, Yusra AI (2012) Carbohydr Polym 87(2):963–979
2. Ajayan PM, Schadler LS, Braun PV (2006) Nanocomposite science and technology. Wiley, New York
3. Andrade JD (1976) Hydrogels for medical and related applications. ACS Publications, Washington
4. Andrieux S, Quell A, Drenckhan W, Stubenrauch C (2018) Adv Colloid Interface Sci 256:276–290
5. Aste T, Weaire D (2008) The pursuit of perfect packing. CRC Press, Boca Raton
6. Barbetta A, Barigelli E, Dentini M (2009) Biomacromolecules 10(8):2328–2337
7. Barbetta A, Carrino A, Costantini M, Dentini M (2010) Soft Matter 6:5213–5224
8. Berger J, Reist M, Mayer J, Felt O, Gurny R (2004) Eur J Pharm Biopharm 57(1):35–52
9. Berger J, Reist M, Mayer J, Felt O, Peppas N, Gurny R (2004) Eur J Pharm Biopharm 57(1):19–34
10. Berghmans J (1973) Chem Eng Sci 28:2005–2011
11. Berglund L (2005) Cellulose-based nanocomposites
12. Berth G, Dautzenberg H, Peter MG (1998) Carbohydr Polym 36(2):205–216
13. Bey H, Wintzenrieth F, Ronsin O, Höhler R, Cohen-Addad S (2017) Soft Matter 13(38):6816–6830
14. Bhattarai N, Gunn J, Zhang M (2010) Adv Drug Deliv Rev 62(1):83–99
15. Binks BP (2002) Curr Opin Colloid Interface Sci 7(1):21–41
16. Blackley DC (2012) Polymer latices: science and technology volume 3: applications of latices. Springer Science & Business Media, Dordrecht

17. Brack H, Tirmizi S, Risen W (1997) Polymer 38(10):2351–2362
18. Butler MF, Ng Y-F, Pudney PDA (2003) J Polym Sci A 41(24):3941–3953
19. Cantat I, Cohen-Addad S, Elias F, Graner F, Höhler R, Pitois O, Rouyer F, Saint-Jalmes A (2013) Foams - structure and dynamics. Oxford University Press, Oxford
20. Chung K-Y, Mishra NC, Wang C-C, Lin F-H, Lin K-H (2009) Biomicrofluidics 3(2):022403
21. Colard CA, Cave RA, Grossiord N, Covington JA, Bon SA (2009) Adv Mater 21(28):2894–2898
22. Costantini M, Colosi C, Guzowski J, Barbetta A, Jaroszewicz J, Święszkowski W, Dentini M, Garstecki P (2014) J Mater Chem B 2(16):2290–2300
23. Costantini M, Colosi C, Jaroszewicz J, Tosato A, Święszkowski W, Dentini M, Garstecki P, Barbetta A (2015) ACS Appl Mater Interfaces 7(42):23660–23671
24. Costantini M, Colosi C, Mozetic P, Jaroszewicz J, Tosato A, Rainer A, Trombetta M, Święszkowski W, Dentini M, Barbetta A (2016) Mater Sci Eng C 62:668–677
25. Croisier F, Jéróme C (2013) Eur Polym J 49(4):780–792
26. Dabo M (2015) Analyse du comportement mécanique des mousses polymères: apport de la tomographie X et de la simulation numérique, Université de Strasbourg, PhD thesis
27. de Gennes P (1979) Scaling concepts in polymer physics. Cornell University Press, Ithaca
28. de Gennes P, Brochard-Wyart F, G. Quéré D (2005) Bulles, Perles et Ondes, 2nd edn. Belin, Paris
29. Delmar K, Bianco-Peled H (2015) Carbohydr Polym 127:28–37
30. Denkov N (2006) "Surfactant Adsoprtion" - Lecture Slides from the Winter School on Foam Physics (Les Houches)
31. Dijkstra M, Hansen JP, Madden P (1995) Phys Rev Lett 75:2236–2239
32. Dimida S, Demitri C, De Benedictis VM, Scalera F, Gervaso F, Sannino A (2015) J Appl Polym Sci 132(28):42256–42264
33. Djerassi C, Gray J, Kincl F (1960) J Org Chem 25(12):2174–2177
34. Dobrynin AV, Colby RH, Rubinstein M (1995) Macromolecules 28(6):1859–1871
35. Doi SFEM (1988) The theory of polymer dynamics. Oxford University Press, Oxford
36. Drenckhan W (2014) From liquid to solid foams. Habilitation à Diriger des Recherches
37. Drenckhan W (2017) Personal communication
38. Drenckhan W, Langevin D (2010) Curr Opin Colloid Interface Sci 15(5):341–358
39. Drenckhan W, Hutzler S (2015) Adv Colloid Interface Sci 224:1–16
40. Drenckhan W, Saint-Jalmes A (2015) Adv Colloid Interface Sci 222:228–259
41. Drury JL, Mooney DJ (2003) Biomaterials 24(24):4337–4351
42. Dukhin SS, Kretzschmar G, Miller R (1995) Dynamics of adsorption at liquid interfaces: theory, experiment, application. Elsevier, Amsterdam
43. Eaves D (2004) Handbook of polymer foams
44. Eichhorn S, Baillie C, Zafeiropoulos N, Mwaikambo L, Ansell M, Dufresne A, Entwistle K, Herrera-Franco P, Escamilla G, Groom L et al (2001) J Mater Sci 36(9):2107–2131
45. Elsing J, Stefanov T, Gilchrist MD, Stubenrauch C (2017) Phys Chem Chem Phys 19:5477–5485
46. Everett DH (1988) Basic principles of colloid science. RSC Publishing, Cambridge
47. Evert R, Eichhorn S (2012) Raven biology of plants. W. H. Freeman, New York
48. Exerowa D, Kruglyakov PM (eds) (1998) Foam and foam films, theory, experiment, application. Elsevier, Amsterdam
49. Fernandes SC, Freire CS, Silvestre AJ, Neto CP, Gandini A, Berglund LA, Salmén L (2010) Carbohydr Polym 81(2):394–401
50. Fernandes SC, Freire CS, Silvestre AJ, Pascoal Neto C, Gandini A (2011) Polym Int 60(6):875–882
51. Gabbrielli R, Meagher AJ, Weaire D, Brakke KA, Hutzler S (2012) Philos Mag Lett 92(1):1–6
52. Gaillard T (2016) Ecoulements confinés à haut et bas Reynolds: génération millifluidique de mousse et drainage de films minces de copolymères, Paris Saclay, PhD thesis
53. Garstecki P, Fuerstman MJ, Stone HA, Whitesides GM (2006) Lab Chip 6:437–446

54. Gibson LJ, Ashby MF (1997) Cellular solids: structure and properties. Cambridge solid state science series. Cambridge University Press, Cambridge
55. Goddard ED, Ananthapadmanabhan KP et al (1993) Interactions of surfactants with polymers and proteins. CRC Press, Boca Raton
56. Goddard E (2002) J Colloid Interface Sci 1:228–235
57. Goddard E, Hannan R (1976) J Colloid Interface Sci 55(1):73–79
58. Guillermic RM, Salonen A, Emile J, Saint-Jalmes A (2009) Soft Matter 5:4975–4982
59. Hinestroza J, Netravali AN (2014) Cellulose based composites: new green nanomaterials. Wiley, New York
60. Höhler R, Sang YYC, Lorenceau E, Cohen-Addad S (2008) Langmuir 24(2):418–425
61. Isert N, Maret G, Aegerter CM (2013) Eur Phys J E 36(10):116
62. Jones RA (2002) Soft condensed matter. Oxford University Press, Oxford
63. Khan H, Shukla R, Bajpai A (2016) Mat Sci Eng C 61:457–465
64. Kimmins SD, Cameron NR (2011) Adv Funct Mater 21(2):211–225
65. Ladet S, David L, Domard A (2008) Nature 452(7183):76–79
66. Landau LD, Lifshitz EM (1959) Fluid mechanics, vol 6. Pergamon Press, Oxford
67. Lau THM, Wong LLC, Lee K-Y, Bismarck A (2014) Green Chem 16:1931–1940
68. Lee S-W, Lim J-M, Bhoo S-H, Paik Y-S, Hahn T-R (2003) Anal Chim Acta 480(2):267–274
69. Li D (2008) Encyclopedia of microfluidics and nanofluidics. Springer, Berlin
70. Li X, Karakashev SI, Evans GM, Stevenson P (2012) Langmuir 28(9):4060–4068
71. Maestro A, Drenckhan W, Rio E, Höhler R (2013) Soft Matter 9:2531–2540
72. Martino AD, Sittinger M, Risbud MV (2005) Biomaterials 26(30):5983–5990
73. Martoïa F, Cochereau T, Dumont P, Orgéas L, Terrien M, Belgacem M (2016) Mater Des 104:376–391
74. Mensire R, Lorenceau E (2017) Adv Colloid Interface Sci
75. Mi F-L, Shyu S-S, Peng C-K (2005) J Polym Sci A 43(10):1985–2000
76. Mills N (2007) Polymer foams handbook: engineering and biomechanics applications and design guide. Elsevier Science & Technology, Oxford
77. Murakami R, Bismarck A (2010) Adv Funct Mater 20(5):732–737
78. Muzzarelli RA (1973) Natural chelating polymers; alginic acid, chitin and chitosan. Pergamon Press, Oxford
79. Muzzarelli RA (2009) Carbohydr Polym 77(1):1–9
80. Oh SH, Park IK, Kim JM, Lee JH (2007) Biomaterials 28(9):1664–1671
81. Orabona E, Caliò A, Rendina I, Stefano LD, Medugno M (2013) Micromachines 4(2):206–214
82. O'Sullivan AC (1997) Cellulose 4(3):173–207
83. Pei A, Butchosa N, Berglund LA, Zhou Q (2013) Soft Matter 9:2047–2055
84. Poulard C, Levannier S, Gryson A, Ranft M, Drenckhan W (2017) Adv Eng Mater 19:2
85. Preston RD (1975) Phys Rep 21(4):183–226
86. Pulko I, Krajnc P (2012) Macromol Rapid Commun 33(20):1731–1746
87. Quell A, Elsing J, Drenckhan W, Stubenrauch C (2015) Adv Eng Mater 17(5):604–609
88. Raj WRP, Sasthav M, Cheung HM (1993) J Appl Polym Sci 49(8):1453–1470
89. Rinaudo M, Milas M, Le Dung P (1993) Int J Biol Macromol 15(5):281–285
90. Rio E, Drenckhan W, Salonen A, Langevin D (2014) Adv Colloid Interface Sci 205:74–86
91. Rojas J, Bedoya M, Ciro Y (2015) Current trends in the production of cellulose nanoparticles and nanocomposites for biomedical applications. In: Cellulose-fundamental aspects and current trends. Intech
92. Roussel J (2015) Mécanique des Fluides, Cours - Compléments. http://www.femto-physique.fr/mecanique_des_fluides/
93. Rubinstein M, Colby RH, Dobrynin AV (1994) Phys Rev Lett 73(20):2776
94. Saint-Jalmes A (2006) Soft Matter 2:836–849
95. Salonen A, Lhermerout R, Rio E, Langevin D, Saint-Jalmes A (2012) Soft Matter 8:699–706
96. Schneider M, Zou Z, Langevin D, Salonen A (2017) Soft Matter

97. Schüler F, Schamel D, Salonen A, Drenckhan W, Gilchrist MD, Stubenrauch C (2012) Angew Chem Int Ed 51(9):2213–2217
98. Sehaqui H, Salajková M, Zhou Q, Berglund LA (2010) Soft Matter 6(8):1824–1832
99. Settles GS, Hargather MJ (2017) Meas Sci Technol 28(4):042001
100. Shen E-C, Wang C, Fu E, Chiang C-Y, Chen T-T, Nieh S (2008) J Periodontal Res 43(6):642–648
101. Shu X, Zhu K (2002) Int J Pharm 233(1):217–225
102. Silverstein MS (2014) Prog Polym Sci 39(1):199–234
103. Silverstein MS (2014) Polymer 55(1):304–320
104. Silverstein MS (2017) Polymer 126:261–282
105. Silverstein MS, Cameron NR (2002) PolyHIPEs — porous polymers from high internal phase emulsions. Encyclopedia of polymer science and technology. Wiley, New York ISBN 9780471440260
106. Silverstein MS, Cameron NR, Hillmyer MA (2011) Porous polymers. Wiley, New York
107. Stubenrauch C, Albouy P-A, Klitzing RV, Langevin D (2000) Langmuir 16(7):3206–3213
108. Stubenrauch C, Von Klitzing R (2003) J Phys Condens Matter 15(27):R1197
109. Svagan AJ, Jensen P, Dvinskikh SV, Furó I, Berglund LA (2010) J Mater Chem 20(32):6646–6654
110. Svagan AJ, Berglund LA, Jensen P (2011) Appl ACS Mater Interfaces 3(5):1411–1417
111. Tabeling P (2005) Introduction to microfluidics. Oxford University Press on Demand, Oxford
112. Testouri A (2012) Highly structures polymer foams from liquid foam templates using millifluidic lab-on-a-chip techniques, Université Paris-Sud XI, PhD thesis
113. Testouri A, Honorez C, Barillec A, Langevin D, Drenckhan W (2010) Macromolecules 43(14):6166–6173
114. Testouri A, Ranft M, Honorez C, Kaabeche N, Ferbitz J, Freidank D, Drenckhan W (2013) Adv Eng Mater 15(11):1086–1098
115. Tsuchida E, Abe K (1982) Interactions between macromolecules in solution and intermacromolecular complexes. Springer, Berlin, pp 1–119
116. van der Net A, Drenckhan W, Weaire D, Hutzler S (2006) Soft Matter 2:129–134
117. van der Net A, Delaney GW, Drenckhan W, Weaire D, Hutzler S (2007) Colloids Surf A 309:117–124
118. van der Net A, Gryson A, Ranft M, Elias F, Stubenrauch C, Drenckhan W (2009) Colloids Surf A 346:5–10
119. Wang Y, Uetani K, S, Zhang X, Wang Y, Lu P, Wei T, Fan Z, Shen J, Yu H, Li S, Zhang Q, Li Q, Fan J, Yang N, Wang Q, Liu Y, Cao J, Li J (2016) ChemNanoMat
120. Weaire D, Phelan R (1994) Philos Mag Lett 69(2):107–110
121. Weaire D, Hutzler S (1999) The physics of foams. Oxford University Press, Oxford
122. Wong JCH, Tervoort E, Busato S, Gonzenbach UT, Studart AR, Ermanni P, Gauckler LJ (2009) J Mater Chem 19:5129–5133
123. Yano H, Nakahara S (2004) J Mater Sci 39(5):1635–1638
124. Yano H, Hirose A, Inaba S (1997) J Mater Sci Lett 16(23):1906–1909
125. Yasunaga K, Neff R, Zhang X, Macosko C (1996) J Cell Plast 32(5):427–448
126. Zhang X, Davis H, Macosko C (1999) J Cell Plast 35(5):458–476
127. Zhang X, Macosko C, Davis H, Nikolov A, Wasan D (1999) J Colloid Interface Sci 215(2):270–279
128. Zimmermann T, Pöhler E, Geiger T (2004) Adv Eng Mater 6(9):754–761

Chapter 3
Preliminary Work: From Liquid to Solid Foams

The study presented in this Thesis was first carried out with the high molecular weight chitosan. The solubility limit of the polymer itself was determined to be 1.5 wt% although some impurities could not be dissolved. These impurities, however, could be removed by filtration under vacuum, using filter paper (#113 from Whatman, with a pore size of 30 µm). Unfortunately, this filtration step caused a polymer loss so that the polymer concentration in the filtrate was no longer precisely known. Because of the filtration step and the low solubility of this polymer, we decided after a year of work to use a purer, low molecular weight chitosan for the main part of the Thesis (Sects. 4 and 5). The work done in the first year was very important for developing a know-how for microfluidic bubbling, and helped set a protocol for the solidification of the liquid templates via several steps of back and forth from liquid to solid foams. Most of the results obtained during the first year are published in [1].

3.1 Chitosan in Solution and Its Gelation

Surface activity Surface tensiometry was used to assess the surface activity of pure chitosan as well as how it interacts with the surfactant Plantacare 2000 UP. The chitosan concentration $c_{chitosan}$ was fixed at 1 wt% and the surfactant concentration $c_{surfactant}$ was increased for the surface tension measurements. The results are shown in Fig. 3.1. We consider as reference the surface tensions of pure water (72.8 mN m^{-1}) and of the AcOH/NaOAc solution (65.8 mN m^{-1}), which consits in 0.05 mol L^{-1} sodium acetate and 0.2 mol L^{-1} acetic acid in water, in which chitosan was dissolved. The surface tension of 1 wt% of chitosan in AcOH/NaOAc is 55.12 mN m^{-1}.

Looking at Fig. 3.1 one notices that the surface tensions of surfactant-containing AcOH/NaOAc solutions are all lower than those containing surfactant and water only. This effect, stronger at low surfactant concentrations, is not surprising considering that the surface tension of the acidic solvent is 65.8 mN m^{-1}[6]. The critical

© Springer Nature Switzerland AG 2019
S. Andrieux, *Monodisperse Highly Ordered and Polydisperse Biobased Solid Foams*,
Springer Theses, https://doi.org/10.1007/978-3-030-27832-8_3

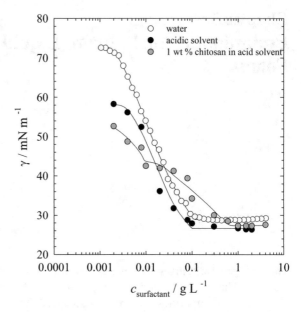

Fig. 3.1 Surface tension γ against the surfactant concentration $c_{\text{surfactant}}$, the surfactant being Plantacare 2000 UP, in water (data from [8]), in the acidic solvent (AcOH/NaOAc), and in the acidic solvent in presence of 1 wt% chitosan (adapted from [7])

micellar concentration (cmc) of Plantacare 2000 UP is the same in water and in the acidic solvent, namely $\sim 0.1\,\mathrm{g\,L}^{-1}$. As regards the surface activity of chitosan, one sees that it is surface active at low surfactant concentrations since the surface tensions are lower in comparison with the chitosan-free solutions. Following the same mechanisms as explained in Sect. 2.5.1, one may conclude that chitosan has a critical aggregation concentration (cac) at $c_{\text{surfactant}} \sim 0.01\,\mathrm{g\,L}^{-1}$. For $c_{\text{surfactant}} \geq$ cac, the chitosan forms complexes with the surfactant molecules—in our case the negatively-charged impurities present in the commercial technical surfactant—and leaves the liquid—air interface in favour of the surfactant. Since the complexation of chitosan molecules in bulk requires surfactant, the critical micellar concentration is shifted to higher values, namely at $c_{\text{surfactant}} \sim 0.7\,\mathrm{g\,L}^{-1}$, as opposed to a cmc at $c_{\text{surfactant}} \sim 0.1\,\mathrm{g\,L}^{-1}$ in the absence of chitosan. Thus, depending on the surfactant content, the chitosan molecules either concentrate at the interface or are dispersed in the bulk. This could be taken advantage of when designing the solid foams as it could provide a way to tailor the foams morphology: according to whether gelation happens preferentially at the interface or in the bulk, one can expect either internal density variations or a material which is homogeneous in its continuous phase. Moreover, if the polymer is strongly anchored at the air-liquid interface, one can assume that the film will be less likely to break and the foam template will preferably yield closed-cell foams, such as for Pickering stabilised foams (see Sect. 2.3.3). We use throughout this Thesis a surfactant concentration of $c_{\text{surfactant}} = 1\,\mathrm{g\,L}^{-1}$ (i.e. 0.1 wt% with respect to the solvent), which means that the chitosan molecules are mainly dispersed in the liquid phase of the foam template.

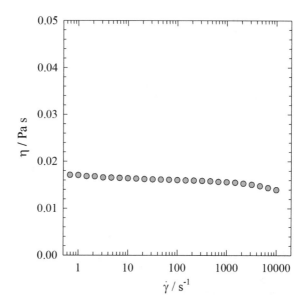

Fig. 3.2 Dynamic viscosity η as a function of the shear rate $\dot{\gamma}$ for the 1.5 wt% chitosan solution in presence of 0.1 wt% surfactant (Plantacare 2000 UP)

Rheological behaviour We assessed the rheological behaviour of the 1.5 wt% chitosan solution with 0.1 wt% surfactant (Plantacare 2000 UP) through the variation of the viscosity η with the shear rate $\dot{\gamma}$ (see Fig. 3.2). The chitosan solution shows a shear thinning behaviour over the whole range of shear rates studied, although shear thinning becomes more important for $\dot{\gamma} > 3000$ s^{-1}. In the literature, the rheological data of chitosan solutions could be approximated by the Cross model, which describes a shear thinning behaviour at high shear rates and takes into account the existence of a Newtonian plateau at low shear rates [4, 9]. However, we see in Fig. 3.2 that although shear thinning is not strong at low shear rates, one cannot speak of a Newtonian plateau either. The absence of a Newtonian plateau may come from the presence of surfactant. Indeed, assuming that the surfactant structures in some way the chitosan molecules in solution, increasing the shear rate can result in the destruction of the said structures and induce a decrease in viscosity.

According to the calculations in Appendix A.2, the flow rates in the chip channels are $\gamma \sim 3$–600 s^{-1} depending on the chip used. The viscosity of the chitosan solution varies thus little around the value $\eta \sim 0.016$ Pa s.

Kinetics of cross-linking We investigated the kinetics of cross-linking as a function of the temperature via oscillatory rheology. Looking at the storage modulus G' and the loss modulus G'' one can determine the time at which the sol-gel transition occurs. Indeed, percolation is reached when the storage modulus becomes higher than the loss modulus, i.e. at the intersection of both curves, as shown in Fig. 3.3a. One sees in Fig. 3.3b that, at constant cross-linker concentration, the gel point t_{gel} is reached earlier as one increases the gelation temperature T_{gel}.

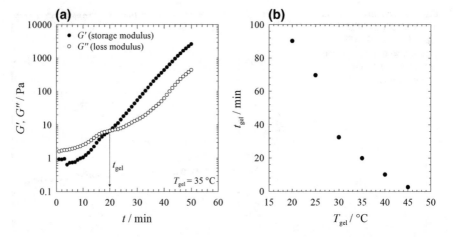

Fig. 3.3 **a** Representative example of a gel point measurement via oscillatory rheometry at 35 °C, for a deformation of 1% and at a frequency of 1 Hz. The arrow shows the gel point t_{gel}, which is the time at which the storage and loss moduli intersect. **b** Gel point as a function of the gelation temperature measured with the same method as in a). The hydrogels were 1.5 wt% chitosan solutions cross-linked with 0.2 wt% genipin and the composition was the same for all temperatures (from [1])

A temperature dependency of the gelation reaction has already been shown by Butler et al., who showed that the kinetics of the cross-linking of chitosan by genipin obeyed an Arrhenius law [3] reading

$$t_{gel} \sim A \exp \frac{-E_A}{R T_{gel}}, \tag{3.1}$$

with A being a constant and E_A the activation energy of the cross-linking reaction. We can determine the activation energy of the cross-linking reaction from an exponential fit, giving $E_A = -1.53 \pm 0.12$ kJ mol^{-1}. Butler et al. [3] found an activation energy equal to -39 kJ mol^{-1} for the system they investigated. The large difference between the activation energy we calculated and the one found in the literature may be attributed to the fact that they used a different chitosan. We know indeed that chitosan, and biobased polymers in general, have different molar masses and deacetylation degrees according to the supplier, and even the batch. The chitosan used by Butler et al. had a DD of 90%, while the chitosan used in the present work has a DD of 80%. However, we show here that the gelation time t_{gel} can be tailored over a wide time range (more than one hour) within a temperature range acceptable for foam stability without having to change the chemistry of the system. Being able to change the gelation time without changing the composition of the system is the best way to adapt the gelation time to the other timescales involes in foam templating.

3.2 Microfluidic Bubbling

Microfluidic chips We used two different microfluidic chips in the preliminary work described here, both with a flow-focusing geometry but with different dimensions (see Fig. 3.4). The first chip is a commercially available glass chip from Dolomite microfluidics with a constant channel depth of 190 μm and will be denoted as the 190 μm chip. The second chip, denoted as the 400 μm chip, is a self-made COC chip, the fabrication process of which is described in Sect. 7.4. The constriction of the 400 μm chip is a square section of 400 μm and a width of 300 μm. The main channels are 1 mm wide and 800 μm deep.

The first foams were prepared from a 1.5 wt% chitosan solution with 0.1 wt% Plantacare 2000 UP. The cross-linker genipin was dissolved directly in the chitosan

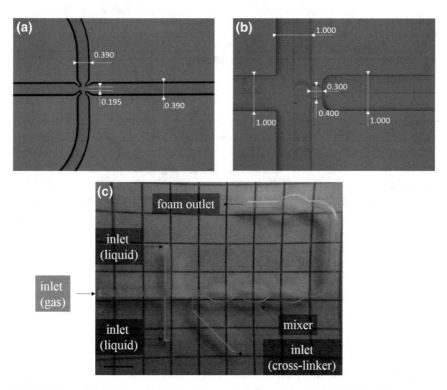

Fig. 3.4 Pictures of the two microfluidic chips used for the studies described in in this chapter. **a** 190 μm chip from Dolomite microfluidics. The channel depth is 0.190 mm. **b** Self-made 400 μm chip. The depth of the channels with a width of 1.000 mm is 0.800 mm while the depth of the constriction, which has a width of 0.400 mm, is 0.400 mm. The dimensions are given in mm. **c** Picture of the master chip for the production of the 400 μm chip not only showing the region around the constriction from **b** but also the fourth inlet available for the addition of cross-linker after bubbling and the mixers designed to mix the polymer solution and the cross-linker solution in the chip. The scale bar is 5 mm (adapted from [1])

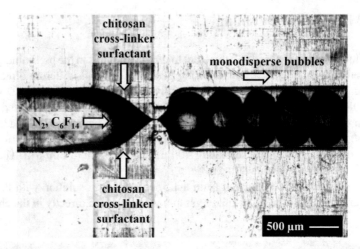

Fig. 3.5 Bubble formation in a COC chip of square constriction (400 μm) of a 1.5 wt% chitosan solution containing 0.2 wt% genipin and 0.1 wt% Plantacare 2000 UP. The pressures applied were p_{gas} = 77 mbar and p_{liquid} = 100 mbar. The gas phase is nitrogen with traces of perfluorohexane C_6F_{14}

solution, which was kept in an ice bath to prevent ill-timed gelation. The highest cross-linker concentration possible without observing an early gelation (*early* meaning before entering the chip or in the chip) was 0.2 wt%. Indeed, for $c_{genipin} > 0.2$ wt% and despite the ice bath, the solution started gelling in the microfluidic chip and clogged the channels. The bubbling was carried out using a pressure pump with two outputs; each pressure could be controlled independently from the other. An example of monodisperse bubbling with the 400 μm chip is shown in Fig. 3.5. Observe the pinch-off of the gas stream (characteristic of the dripping regime) resulting in the formation of a bubble, as explained in Sect. 2.4.

This microfluidic bubbling procedure allowed us to generate monodisperse bubbles in a reproducible manner. However, at first, in order to prevent the clogging of the channels, we had designed a fourth inlet in the chip (after the constriction, see Sect. 7.4) for the separate addition of the cross-linker solution, as shown in Fig. 3.4c. This method was already used by Testouri et al. for the cross-linking of chitosan with glyoxal: adding the cross-linker at a later stage in the microfluidic set-up prevented the chitosan solution from gelling while still in the channels [9, 10]. However, we decided against using this inlet channel. Indeed, in order to get monodisperse bubbles, the pressure has to remain constant with time at every point of the microfluidic set-up, which implies the stability of the flows at every point in the chip. However, for a too low flow rate of the genipin solution (genipin being the cross-linker used in this Thesis), the bubbly flow entered the channel which served as an inlet for the cross-linker solution, until it was pushed back in the main channel by the incoming cross-linker solution. This resulted in flow instabilities which prevented us from reaching monodispersity. A higher flow rate of the cross-linker solution may pre-

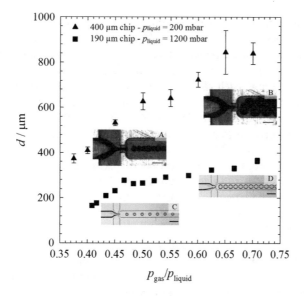

Fig. 3.6 Bubble diameter d plotted versus the ratio of gas and liquid pressure p_{gas}/p_{liquid} for the 190 μm and the 400 μm microfluidic chips. The liquid pressure is constant in each chip and equal to 200 mbar for the 400 μm chip and 1200 mbar for the 190 μm chip. The error bars represent the standard deviation of the bubble sizes of the collected foams and are smaller than the symbols for the 190 μm chip. The photographs were taken during bubble generation. The bubbles shown are formed at gas pressures p_{gas} of **a** 75 mbar and **b** 110 mbar for the 400 mm chip, and **c** 520 mbar and **d** 800 mbar for the 190 μm chip. The scale bars are 500 μm. Figure taken from [1]

vent the entry of the chitosan solution in the cross-linker inlet, and thus help restore monodispersity. However, increasing the cross-linker flow rate would result in an unwanted dilution of both the chitosan solution and the genipin solution. Therefore, we decided against using this fouth inlet channel and dissolved the cross-linker directly in the chitosan solution instead.

Chip calibration In order to find out which pressures yield which bubble sizes, one needs to calibrate the microfluidic chips, i.e. measure the bubble size as a function of the pressure ratio p_{gas}/p_{liquid}, as shown in Fig. 3.6. We fixed for each chip the liquid pressure p_{liquid} and varied the gas pressure p_{gas} over the bubbling range. The bubbling range is the pressure ratio range over which bubbling occurs. For a too low pressure ratio, the liquid flows into the gas channel and no bubble is produced, whereas for a too high pressure ratio, the gas goes into the liquid channels and no bubble is produced. The flow of gas into the liquid channels does not constitute a problem in itself, but one must avoid the flow of liquid into the gas channel, especially if one is handling a polymer solution with cross-linker. Indeed, if the channel is wetted by the polymer solution, even shortly, the polymer cross-links and dries into a solid layer difficult to clean away, modifying the dimensions of the channels and even clogging the chip.

One sees in Fig. 3.6 that by setting only one liquid pressure for each chip, one has access to bubble sizes between ca. 165 and 845 μm, offering a wide range of accessible pore sizes for the polymer foams made out of the liquid foam templates. At a constant liquid pressure, the bubble size increases with increasing gas pressure p_{gas}. Looking closer at each chip, one sees that the bubble sizes accessible with the 190 μm chip range from 165 to 365 μm and the bubble sizes accessible with the 400 μm range from 370 to 845 μm. Note that the 190 μm chip yields foams with a very low polydispersity (the error bars are smaller than the symbols), while the standard deviations are larger for the 400 μm chip. We ascribe this observation to the larger dimensions of the 400 μm in comparison to the 190 μm chip, which implies that smaller pressures are required to push the fluids in the 400 μm than in the 190 μm chip. The larger polydispersities for the foams made with the 400 μm chip comes from the fact that the pressure controller is more subject to fluctuations at low pressures [1].

Note that changing the pressure ratio p_{gas}/p_{liquid} allows to tune the bubble size but also the liquid fraction of the foam out of the microfluidic channel. We did not, however, measure the liquid fractions for each pressure ratio p_{gas}/p_{liquid} since the foams are subject to drainage, which means that the liquid fraction set by the microfluidic bubbling is not equal to the liquid fraction of the liquid foam, as explained in Sect. 2.3.4.

3.3 Gelation of Monodisperse Liquid Chitosan Foams

Monodisperse liquid chitosan foams The monodisperse foams were collected in Petri dishes and left to self-order into hexagonally close-packed structures, as shown in Fig. 3.7 for two monodisperse foams made with the 190 μm chip and the 400 μm chip. The foam made using the 190 μm chip has a bubble size of $d = 338 \pm 8$ μm and the foam made using the 400 μm chip has a bubble size of $d = 644 \pm 30$ μm. Both foams had thus *PDIs* < 5%, namely 1.9% and 4.7%, respectively. The higher *PDI* of the foam made using the 400 μm chip is not surprising considering the larger error bars in the chip calibration diagram for the 400 μm chip (see Fig. 3.6).

The aim of foam templating is to maintain the order and foam structure throughout cross-linking and drying. One has thus to fight against the different foam destabil-isation mechanisms which are evaporation, drainage, coarsening and coalescence (Sect. 2.1.2). An instant solidification of the liquid template would solve all the stability issues at once, but the foams need time to order and cross-link. Although a gelled foam can be considered stable, one has to ensure against the destabilisation of the non-gelled foam until its gelation.

Gelation and drying of monodisperse liquid chitosan foams Figure 3.8 shows a monodisperse chitosan foam which cross-linked 18 h at room temperature. Looking from below (Fig. 3.8a) one sees that the foam retained its monodispersity and hexagonal close-packing throughout cross-linking. However, looking at the top of

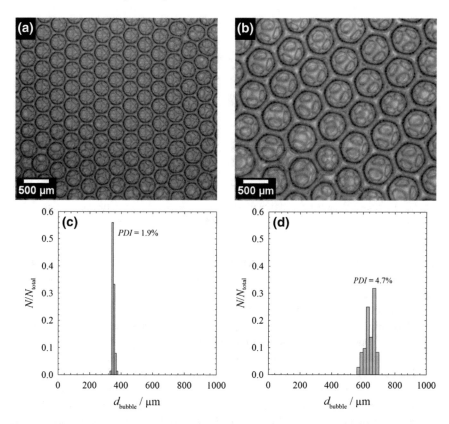

Fig. 3.7 Monodisperse liquid foams prepared using the **a** 190 μm chip and **b** 400 μm chip, and the corresponding bubble size distributions, (**c**) and (**d**). The bubble size distributions are 338 ± 8 μm for the 190 μm chip and 644 ± 30 μm for the 400 μm chip. The pictures were taken shortly after collecting the foams (adapted from [1])

the sample (Fig. 3.8b), one clearly sees that the top layer has been reduced to smaller bubbles sitting on top of a layer of deformed bubbles, with a higher polydispersity than at the bottom of the foam. However, the top layer is not dried and one still has a foamed hydrogel, indicating that evaporation is not the main mechanism at play here. We thus ascribe this observation to gas loss to the atmosphere.

Indeed, the atmosphere can be seen as a bubble of infinite diameter, to which the bubbles in the foam lose their gas through diffusion driven by the capillary pressure of bubbles [5]. On the one hand, gas diffusion to the atmosphere results in a diminution of the total amount of gas present in the foam. On the other hand, Ostwald ripening, i.e. gas diffusion within the foam, results in a change of the bubble size distribution but the overall foam volume is not affected. To confirm this assumption that gas diffusion to the atmosphere is responsible for damaging the foam's top layer, we studied the evolution with time of monodisperse chitosan monolayers in the presence of cross-linker (Fig. 3.9), in order to find out if gelation can stop gas diffusion and/or Ostwald

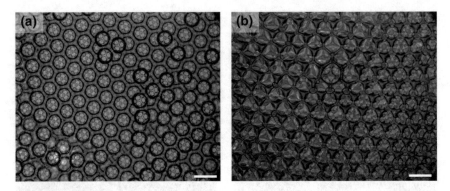

Fig. 3.8 Pictures of the **a** bottom and the **b** top of a monodisperse chitosan foam cross-linked with genipin for 18 h at room temperature in a humid atmosphere. The scale bars are 1 mm

ripening [2, 5]. We studied two different monolayers, one without perfluorohexane in the bubbles and one with perfluorohexane in the bubbles to hinder Ostwald ripening (see Sect. 2.1.2).

Let us first look at the monolayer with bubbles not stabilised against coarsening by C_6F_{14} (Fig. 3.9a–e). The monolayer presented a large number of small bubbles at early times,[1] but they disappeared rapidly due to Ostwald ripening and gas diffusion to the atmosphere as a consequence of their low radii and thus high capillary pressures (see Sect. 2.1.2). We thus excluded them from the calculations of the average bubble size $<d>$ and *PDI* shown in Figs. 3.9k and l. One sees that in the absence of C_6F_{14} the average bubble size decreases with time, although slower at longer times, indicating that the loss of gas from the bubbles to the atmosphere plays an important role. Indeed, one sees that all the bubbles get smaller and no bubble gets bigger, which indicates that gas diffusion to the atmosphere predominates over Ostwald ripening. Looking at the evolution of the *PDI* with time, one sees a strong increase from 5% to 13% within 90 min, which can be attributed to Ostwald ripening. However, the *PDI* decreases between 90 min and 120 min, which arises from the dissolution of the smallest bubbles. Note that the decrease in bubble size, i.e. the gas loss to the atmosphere, is less important between 90 and 120 min than at shorter times (Fig. 3.9k), indicating that cross-linking prevents gas diffusion. Indeed, Bey et al. [2] showed that gelation may induce an arrest of gas diffusion if the elastic modulus of the continuous phase becomes high enough to overcome bubble shinkage. The overall decrease of the average bubble size shows that gas loss to the atmosphere is not negligible, but neither is Ostwald ripening. To sum up, gas diffusion to the atmosphere is mainly responsible for the decrease in overall bubble size observed in Fig. 3.9k, whereas Ostwald ripening is mainly responsible for the strong increase in *PDI* shown in Fig. 3.9l. Both mechanisms are thus at play to destabilise the foam monolayer.

---

[1] Often generated at tubing intersections, these smaller bubbles are no longer formed once the fluidic system has been running long enough.

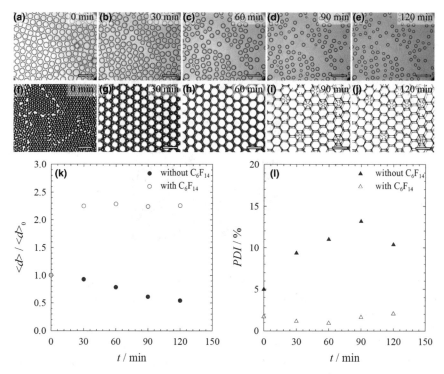

Fig. 3.9 Ageing foam monolayers without and with perfluorohexane (C_6F_{14}). Pictures from **a** to **e** show the same spot of a monodisperse foam monolayer with sole nitrogen as the gas phase from 0 to 120 min after its collection. Pictures from **f** to **i** show the same spot of a monodisperse foam monolayer with nitrogen containing traces of C_6F_{14} as the gas phase from 0 to 120 min after its collection. The scale bars are 1 mm. **k** Evolution of the average bubble size $<d>$ scaled with the initial bubble size $<d>_0$ as a function of time for the monolayers with and without C_6F_{14}. The initial average bubble sizes $<d>_0$ are 436 and 286 μm for the C_6F_{14}-free monolayer and the C_6F_{14}-containing monolayer, respectively. **l** Evolution of the PDI with time for the monolayers with and without C_6F_{14}

We reproduced the experiment with a chitosan foam monolayer stabilised against coar-sening by C_6F_{14} (Fig. 3.9 f–j) and monitored the evolution of the average bubble size and PDI as a function of time (Fig. 3.9k, l). Interestingly, the average bubble size more than doubles within 30 min to remain constant at longer times, without significant increase in polydispersity. The bubbles go from a spherical shape to close-packed hexagons, which corresponds to a two-dimensional jamming transition. The liquid fraction decreases as the bubbles get bigger. Notice as well how some films rupture as early as after 90 min. The presence of perfluorohexane sets a chemical potential in all bubbles counteracting the Laplace pressure difference between the bubbles thus preventing Ostwald ripening. However, adding C_6F_{14} into the gas phase induces a swelling of the bubbles due to a gas intake from the atmosphere. This phenomenon comes directly from the low sol-

Fig. 3.10 Picture of a cross-linked monodisperse chitosan foam stabilised against Ostwald ripening with C_6F_{14}, showing the upper part of the foam with bubbles swollen due to gas intake from the atmosphere, and the lower part of the foam with unswollen bubbles. The red dashed line shows the junction between the two regions. The vial is sealed with a screw cap. The scale bar is 5 mm

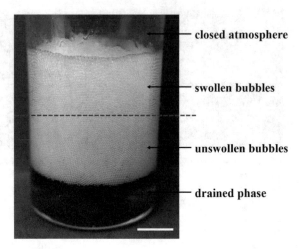

ubility of C_6F_{14} in water and lies on the same physics that is allowing C_6F_{14} to prevent coarsening. Indeed, C_6F_{14} cannot migrate from a bubble to another one and its presence sets an equal chemical potential in all the bubbles. This prevents gas transport between bubbles, i.e. coarsening. However, there is a strong potential difference between the inside of the bubbles and the surrounding atmosphere. Since C_6F_{14} cannot migrate *out* to the atmosphere, this chemical difference is compensated by a migration of atmospheric gas *into* the bubbles, inducing a swelling of the bubbles, as observed in Fig. 3.9. In the case of a bulk foam, the top layer is the most sensitive to this effect, but it can propagate to a few layers below since a swelling of the top layers induces a chemical potential difference between the top layer and the second layer, as seen in Fig. 3.10. One clearly sees two regions, the upper region has swollen ordered bubbles while the lower region has unswollen monodisperse and ordered bubbles. The dashed red line shows the junction between these two regions. One can limit the number of layers with swollen bubbles by diminishing the volume of atmosphere available, i.e. by letting as little space as possible between the top of the foam and the cap of the vial.

We have seen in Sect. 3.1 that the cross-linking of chitosan by genipin can be sped up by increasing the gelation temperature. Figure 3.11 shows how, from a same liquid foam template (Fig. 3.11a), one may obtain different foams with a different gelation history after 20 h.

When left to gel at room temperature, the foam loses its monodispersity, as seen in Fig. 3.11b, as coarsening occurs before gelation can prevent it. Figure 3.11c shows a foam which had been heated up to 40 °C for 2 h after its formation. This gelled foam shows a high crystallinity with distinct grains and grain boundaries. The *PDI*s of the foam left at room temperature for 20 h and of the one heated at 40 °C for 2 h are 16.2 and 4.3%, respectively. In other words, the earlier the gelation, the better the foam's monodispersity and crystallinity can be transferred to the solid counterpart. However, such gelled foams are soft materials, containing a high amount of water, which limits

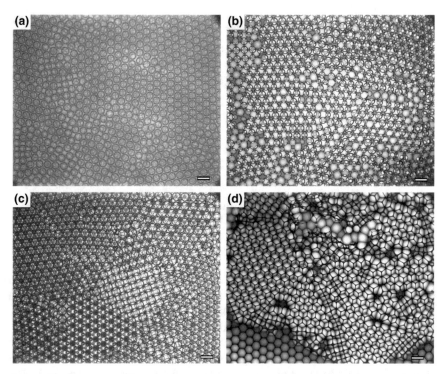

Fig. 3.11 Pictures of monodisperse chitosan foams **a** 5 min after being collected at RT, **b** 20 h after being collected at RT **c** after 2 h at 40 °C and 18 h at RT and **d** after 2 h at 40 °C and 18 h at 60 °C. The monodisperse liquid foams were generated using the 400 μm chip and no C_6F_{14} was used. The scale bars are 1 mm

their range of applications. Such hydrogels can thus not be called macroporous chitosan, but have to be dried if one seeks to obtain monodisperse solid chitosan foams.

Hence we generated a monodisperse hydrogel foam by heating the foam to 40 °C for 2 h, but instead of storing it at room temperature, it was heated up to 60 °C to accelerate evaporation. The resulting structure is shown in Figs. 3.11d and 3.12b. For comparison, the undried monodisperse gelled foam is shown in Fig. 3.12a. The dried foam presents polyhedral pores characteristic of low-density foams, as opposed to its hydrogel counterpart. The transition from spherical bubbles in the hydrogel state to polyhedral pores in the dry state originates form the loss matter whilst drying. Indeed, bear in mind that the foaming solution is 1.5 wt% chitosan in an aqueous solvent. Therefore, the foam has to lose ca. 98.5% of its mass to go from the liquid to the dry state. Despite this consequent mass loss, one can still observe different crystalline structures such as Kelvin cells and defects of cubic geometry between two neighbouring crystalline grains. Although the gelled foams presents a closed-cell structure similar to that of a liquid wet foam, the dried foam is an open-cell foam. The edges of pore walls can be observed at the vertices, as pointed by the red arrows in Fig. 3.12b, indicating that the walls disappeared during drying.

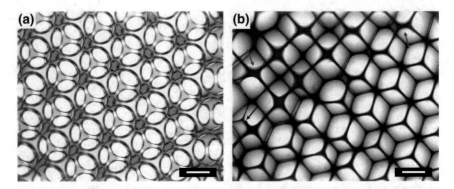

Fig. 3.12 Pictures showing **a** the closed-cell structure of a foam gelled for 2 h at 40 °C and **b** the open-cell structure of a foam gelled for 2 h at 40 °C and dried at 60 °C for 18 h. The scale bars are 500 μm

The dried foam also shows defects consisting of broken Plateau borders (green arrow in Fig. 3.12b), indicating that these regions had been exposed to stresses high enough to break the film. We ascribe such high stresses to the slow drying process, allowing for a reorganisation of the chitosan network as a response to the loss of solvent. The shrinkage of the polymer network may have lead to the low-density structure observed and caused the high stresses responsible for the break-off of the films. Accelerating drying to prevent the hydrogel from compensating the loss of its solvent would allow for a better conservation of the template's structure and avoid internal stresses. A too high temperature is, however, not recommended, as it would harm the foam template's stability. We thus tried freeze-drying, which consists in freezing the foam followed by the sublimation of the ice cristals under vacuum.

3.4   Solidification of Liquid Chitosan Foams

We chose to avoid heating as a drying method because of the damages it causes to the foam structure and turned instead to freeze-drying. The first step of freeze-drying is freezing the foam, which arrests its structure. Since we aim for solid polymer foams, the use of freeze-drying allows us to go directly from a liquid foam to a solid foam without the constraint of having to cross-link chitosan. Figure 3.13 shows solid foams which were freeze-dried at different stages, i.e. before cross-linking (Fig. 3.13a) and after 20 h cross-linking (Fig. 3.13c). The corresponding bubble/pore size distributions in Fig. 3.13b and d show that although starting from liquid templates with comparable bubble size distributions, the resulting solid chitosan foams differ much as to their morphologies and pore size distributions.

The average pore size of the non-cross-linked foam is 402 ± 32 μm and lies ca. 200 μm below the bubble size of the liquid template, which is 644 ± 30 μm. Moreover, the shape of the pores' cut suggest spherical pores with small interconnects. We must,

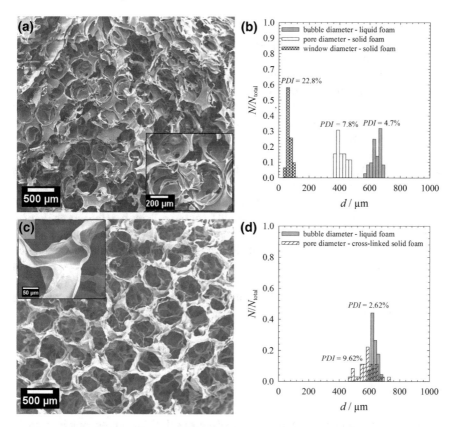

Fig. 3.13 Pictures of monodisperse solid chitosan foams solidified by freeze-drying **a** before cross-linking and **c** after 20 h cross-linking (2 h at 40 °C followed by 18 h at room temperature), with **b** and **d** the pore sizes and bubbles sizes of their respective liquid templates, respectively. The monodisperse liquid foams were generated using the 400 μm chip. Adapted from [1]

however, point out that the pore size distribution as measured from SEM pictures underestimates the actual pore size distribution in bulk as the sample is randomly cut, i.e. the pores are not cut at their hemispheres and the observed diameter is an underestimation of the pore diameter. Similarly, the *PDI* is also overestimated by this measuring method. For the cross-linked solid foam shown in Fig. 3.13c one does not observe a significant difference between the bubble size distribution of the liquid template (630 ±17 μm) and the pore size distribution of the solid foam (591 ± 57 μm).

Moreover, the pores are not spherical and fully open, so that one cannot define pore openings; the foam is solely composed of struts. The struts are, however, strongly deformed and recall the morphology of Plateau borders (see inset in Fig. 3.13c). This morphology may explain why the pore size distribution of the solid foam is close to the bubble size distribution of its liquid counterpart, as one has access to the whole pore from the SEM picture, as opposed to the foam in Fig. 3.13a.

Cross-linking thus plays a huge role in setting the morphology of the solid foam. Although being pure speculation, a couple of explanations can be given. The first possible explanation for the morphological difference between the cross-linked foam and non-cross-linked foam is that the cross-linked foam has a lower liquid fraction due to drainage and the partial evaporation of water occuring during cross-linking, whereas we froze the non-cross-linked foam right after its formation. Moreover, the internal stresses within foams of different densities may induce different types of deformations. Another explanation may be that cross-linking induces internal stresses in the hydrogel foam which result in a deformation of the struts even before freeze-drying; this morphological change is then transposed to the solid foam. A possible way to prove if the hypothesis of a low liquid fraction explains best the morphological difference between the cross-link and non-cross-link foams is to generate cross-linked foams over a wide range of bubble sizes in order to change the liquid fraction of the foams and verify if a lower liquid fraction, i.e. for higher bubble sizes, results in larger Plateau borders. One could also study how internal stresses caused by cross-linking affect foam morphology by changing the cross-linker concentration, i.e. the cross-linker density, all other parameters being kept equal, and study the morphology of the resulting solid foams.

3.5   Conclusion

We have seen that one can obtain a monodisperse solid chitosan foam with various morphologies by modifying the foam templating route (see Fig. 3.14). We can play with the different stages of the foam templating process to tailor the structure of the resulting solid foam. Microfluidic allows generating monodisperse foams with a wide range of average bubble sizes, from ca. 200 to 800 μm. Once the liquid foam template has been produced, it needs to be solidified. We showed that the drying procedure and whether the foam is cross-linked or not strongly affects the morphology of the resulting solid foam. Indeed, as the blue path in Fig. 3.14 shows it, we can obtain two strongly different morphologies from the same cross-linked liquid foam template, depending on the drying process. This is one example amongst others developed in the present chapter. However, after having tested different solidification protocols, we had to find a reproducible protocol which accounts for the different destabilisation mechanisms of liquid foams and yields monodisperse highly ordered solid chitosan foams.

We thus designed the following solidification protocol. We use perfluorohexane during microfluidic bubbling to stabilise the foams against coarsening. We fill the Petri dish up to the top with foam and seal it with Parafilm so that the foam is not in direct contact with the atmosphere. By doing so we prevent the swelling of the

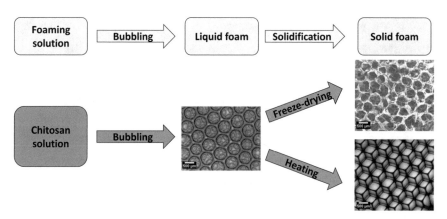

Fig. 3.14 How foam templating can be used to produce monodisperse solid foams with a variety of morphologies. In grey is the general foam templating route towards solid foams. In blue are the variations from this general route leading to various structures. One can vary the pore morphology by changing the drying procedure which follows cross-linking, i.e. either using freeze-drying or heating in an oven

upper layers. We then let the foams cross-link at room temperature for 18 h. Once cross-linking is finished, the foam is frozen in liquid nitrogen and subsequently freeze-dried, yielding a monodisperse solid chitosan foam.

References

1. Andrieux S, Drenckhan W, Stubenrauch C (2017) Polymer 126:425–431
2. Bey H, Wintzenrieth F, Ronsin O, Höhler R, Cohen-Addad S (2017) Soft Matter 13(38):6816–6830
3. Butler MF, Ng Y-F, Pudney PDA (2003) J Polym Sci A 41(24):3941–3953
4. Calero N, Muñoz J, Ramírez P, Guerrero A (2010) Food Hydrocoll 24(6–7):659–666
5. Dutta A, Chengara A, Nikolov A, Wasan D, Chen K, Campbell B (2004) J Food Eng 62(2):177–184
6. Ikeda H, Suzuki A (1995) Ind Eng Chem Res 34(11):4110–4117
7. Schad T (2015) Rheologie and surface tension of chitosan in solution in presence and in absence of surfactant
8. Stubenrauch C (2014) Personal communication
9. Testouri A, Honorez C, Barillec A, Langevin D, Drenckhan W (2010) Macromolecules 43(14):6166–6173
10. Testouri A, Arriaga L, Honorez C, Ranft M, Rodrigues J, van der Net A, Lecchi A, Salonen A, Rio E, Guillermic R-M, Langevin D, Drenckhan W (2012) Colloids Surf A 413:17–24

Chapter 4
Monodisperse and Polydisperse Chitosan Foams via Microfluidics

4.1 Chitosan Solutions and Microfluidic Bubbling

Chitosan solutions and their surface tensions Foam templating offers the possibility to vary many parameters such as the chemical composition of the liquid foam, the foaming conditions or the solidification procedure. We thus had to select the parameters of interest for the study at hand, and chose not to vary the composition of the liquid foam. After a long phase of preliminary work with an inconvenient high molecular weight chitosan, which we describe in detail in Sect. 3, we turned to a low molecular weight chitosan (see Sect. 7.1), which one can dissolve at a concentration of $c_{chitosan} = 4$ wt% in 1 vol% acetic acid. We also kept the surfactant concentration ($c_{surfactant} = 0.1$ wt% $= 1$ g L^{-1}) and the cross-linker concentration ($c_{genipin} = 1$ wt%) constant.

Firstly, we studied the surface tension of chitosan solutions with a constant chitosan concentration $c_{chitosan} = 4$ wt% as a function of the surfactant concentration $c_{surfactant}$.[1] The data are shown in Fig. 4.1. One sees that in the absence of chitosan the cmc of the surfactant Plantacare 2000 UP in 1 vol% acetic acid is the same as in water, namely 0.1 g L^{-1}. In the absence of surfactant the surface tension of 4 wt% chitosan in 1 vol% acetic acid is 37.7 mN m^{-1}. In the presence of surfactant the surface tension γ of the chitosan solution does not vary at low surfactant concentrations. However, for $c_{surfactant} > 0.02$ g L^{-1} the surface tension of the chitosan solution decreases until the cmc is reached at ca. 2 g L^{-1} with a surface tension of 27.6 mN m^{-1}. Unlike for the high molecular weight chitosan (see Fig. 3.1 in Sect. 3.1), one does not observe any cac. Note that the model of the surface activity of a polycation in the presence

---

Parts of this chapter have been reproduced with kind permission from [2]. Copyright (2018) American Chemical Society.

---

[1] A large part of the data presented in this paragraph are the result of experiments carried out by Tamara Schad during her research internship.

© Springer Nature Switzerland AG 2019
S. Andrieux, *Monodisperse Highly Ordered and Polydisperse Biobased Solid Foams*,
Springer Theses, https://doi.org/10.1007/978-3-030-27832-8_4

Fig. 4.1 Surface tension γ of 1 vol% acetic acid and 4 wt% chitosan in 1 vol% acetic acid as a function of the surfactant concentration at 23 °C, the surfactant being Plantacare 2000 UP. The surface tension of the pure 1 vol% acetic acid solution is 68.9 mN m^{-1} and 37.7 mN m^{-1} for the 4 wt% chitosan in 1 vol% acetic acid solution. For comparison, the surface tension of the surfactant in pure water is plotted as a function of the surfactant concentration. Figure adapted from [21]

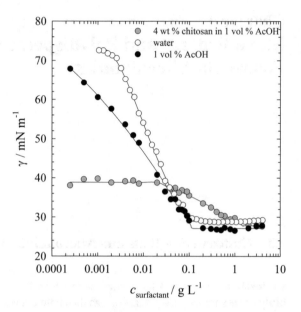

of a negatively charged surfactant presented in Sect. 2.5.1 only partially applies here since the surfactant is not per se anionic. Indeed, the surfactant molecule itself is neutral, but since Plantacare 2000 UP is a technical surfactant, it contains negatively charged surface-active impurities that may lead to the formation of complexes as described in Sect. 2.5.1. There are two possible explanations for the absence of a cac in the surface tension curve: (1) The negatively charged impurities do not suffice to build complexes large enough to induce precipitation, or aggregation occurs at surfactant concentrations higher than the cmc. In that case, the polymer chains at the air-liquid interface are simply replaced by the surfactant molecules in a continuous way. (2) The cac is so close to the cmc that it is not measurable in the present case.

We did not observe any phase separation or precipitation as we measured the surface tension, whereas when washing the glassware and microfluidic chips with dishwashing liquid—mainly composed of negatively-charged sodium dodecyl sulfate—we observed the formation of a white precipitate. It is thus very likely that the first explanation holds true. However, only more accurate experiments such as, e.g., dynamic light scattering, would allow for a definite characterisation of the composition of the bulk phase—and thus the air-liquid interface—and of the interactions between the chitosan molecules and the surfactant.

Rheological behaviour We assessed the rheological behaviour of the 4 wt% chitosan solution with 0.1 wt% surfactant (Plantacare 2000 UP) through the variation of the viscosity with the shear rate (see Fig. 4.2). The chitosan solution shows a Newtonian plateau for shear rates $\dot{\gamma} < 1000$ s^{-1} with a viscosity $\eta = 0.04$ Pa s. The chitosan solution shows a flow focusing behaviour shown by the decrease in viscosity for $\dot{\gamma} > 1000$ s^{-1}. The Newtonian plateau followed by shear thinning at high shear rates is characteristic of the Cross model, which was already observed for chitosan solutions [9, 23].

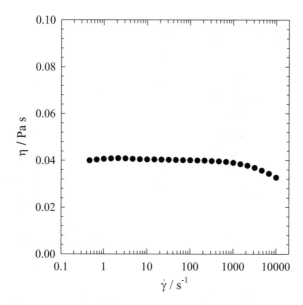

Fig. 4.2 Viscosity η as a function of the shear stress $\dot{\gamma}$ of the 4 wt% chitosan in 1 vol% acetic acid solution in presence of 0.1 wt% surfactant (Plantacare 2000 UP)

We know from the calculations in Appendix A.2 that the flow rate of the chitosan solution in the channels before the cross-junction (flows with a flow rate $1/2 \cdot Q_{chitosan}$ in Fig. 4.3) is $\gamma \sim 210 \, s^{-1}$. The viscosity of the chitosan solution in the chip is thus $\eta \sim 0.04$ Pa s and we show in Appendix A.2 that the flows are laminar.

Microfluidic bubbling: general set-up The principle is based on a microfluidic bubbling method allowing for the generation of foams with controlled polydispersities. Figure 4.3 shows the set-up used throughout this chapter. The microfluidic chip is the 190 μm chip presented in Sect. 3.5 with a cross-flow geometry (see Fig. 3.4a).

The chitosan solution is introduced into the chip with a syringe pump at a constant flow rate $Q_{chitosan} = 180 \, \mu L \, min^{-1}$. We recall that $Q_{chitosan}$ is kept constant for all experiments and at all times. The composition of the chitosan solution is also kept constant with a chitosan concentration of $c_{chitosan} = 4$ wt% and a Plantacare 2000 UP concentration of $c_{surfactant} = 0.1$ wt%. The chitosan flow is split into two flows with a flow rate of $1/2 \cdot Q_{chitosan}$ which meet at the cross-section along with the gas flow, resulting in the formation of bubbles, as explained in Sect. 2.4. Once formed, the bubbly flows continues its way in the main microfluidic channel, and, although the liquid flow rate is once again equal to $Q_{chitosan} = 180 \, \mu L \, min^{-1}$, the flow rate of the bubbly flow is higher due to the presence of the gas bubbles. The genipin solution, which has a genipin concentration of $c_{genipin} = 1$ wt% and a Plantacare 2000 UP concentration of $c_{surfactant} = 0.1$ wt%, is added to this bubbly flow at a flow rate of $Q_{genipin} = 1/6 \cdot Q_{chitosan} = 30 \, \mu L \, min^{-1}$. The genipin flow rate $Q_{genipin}$ is also kept constant so that the cross-linking degree remains constant regardless of the gas

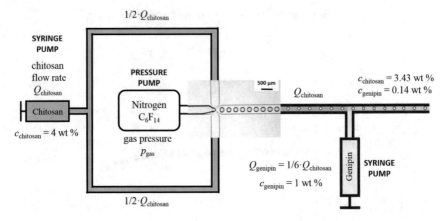

Fig. 4.3 Generation of monodisperse and polydisperse chitosan foams using a microfluidic device. Figure taken from [2]

pressure p_{gas}. When mixing the chitosan and the genipin solutions both solutions are diluted, yielding final concentrations of 3.43 wt% chitosan, 0.14 wt% genipin and 0.1 wt.% surfactant. This composition does not change throughout this chapter.

The gas flow is monitored by means of a pressure controller which is connected to the nitrogen gas system. We set a glass bottle between the pressure controller and the chip which contains liquid perfluorohexane C_6F_{14} to prevent Ostwald ripening (see Sect. 3.3). Traces of perfluorohexane are taken away with the nitrogen as the gas flows in and out of the bottle. The gas phase is thus composed of a mixture of nitrogen and C_6F_{14}. The gas flow rate is controlled by changing the pressure but cannot be quantitatively measured. The pressure, however, can be precisely tuned, which allows for the variation of bubble sizes, as shown by the calibration of the microfluidic chip (see Fig. 4.4).

Microfluidic bubbling for monodisperse foams Every other parameter being kept constant, one can easily fine-tune the bubble size by controlling the gas pressure p_{gas} on the pressure controller, as shown in Fig. 4.4. One observes a clear increase of the bubble size with increasing gas pressure, confirming the trends observed during the preliminary studies in Sect. 3.1. One sees a lower region of gas pressure where the increase in bubble size is steep, followed by a larger region, at higher gas pressures, where the increase in bubble size is much slower. The gas pressure p_{gas} at which the slope changes lies between 600 and 700 mbar. Bubble sizes from 180 μm to 400 μm are accessible with the used chip. We plot in Fig. 4.4 the diameter of the bubbles once they left the microfluidic set-up (grey circles) as well as the bubble diameter in the microfluidic channel (black triangles). The bubble size measured in the chip is systematically smaller than the bubble size outside of the microfluidic set-up. The latter is the bubble size that one needs to measure when talking about the bubble size of the foam template. When measured in the chip, the bubble size reaches an upper

Fig. 4.4 Bubble diameter d_{bubble} measured inside and outside the chip as a function of the gas pressure p_{gas}. The chitosan flow rate was set to $Q_{chitosan} = 180\ \mu L\ min^{-1}$. The insets show pictures of the microfluidic bubbling (C, D) and the corresponding foam monolayers (A, B) from which the bubble sizes were determined at $p_{gas} = 500$ mbar (A, C) and $p_{gas} = 1300$ mbar (B, D). The error bars correspond to the standard deviations. All scale bars are 500 μm. Figure taken from [1]

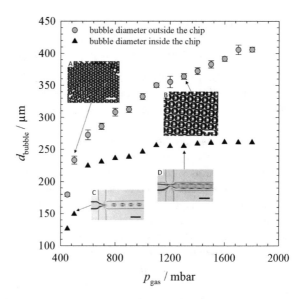

limit at ca. 260 μm, which is smaller than the width of the channel (equal to 390 μm) and bigger than the height of the channel (equal to 190 μm). We refer to Fig. 3.4a for more details.

However, once outside the microfluidic chip, the bubbles can reach diameters up to 400 μm, i.e. larger than the channel's height, which confirms that the bubbles are flattened within the channel. We refer to van der Net et al. [26] where it is explained why the bubbles in the microfluidic channel are smaller than outside the chip. Firstly, optical effects and light refraction in the glass channels cause an underestimation of the actual bubble diameter. Secondly, the bubble diameter was determined from the height of the bubbles since this is the only dimension that can be measured at all gas pressures. Indeed, as one can see in Fig. 4.4c, the high-speed camera is not always able to capture clear images of the bubbles which appear no longer spherical but elongated in the direction of the flow. Thirdly, at higher gas pressures, the bubble stream appears as a continuous train of bubbles (see Fig. 4.4d), making it impossible to observe the limits of the bubbles. By measuring the bubble sizes in the direction perpendicular to the flow, we neglected the confinement of the bubbles within the channels and the bubble deformation due to the viscous stresses from the chitosan solution. After this demonstration of the inaccuracy of the measurement of the bubble size within the microfluidic channel, the reader may wonder why we measured the bubble size within the microfluidic channel. Firstly, the comparison of the bubble sizes inside and outside of the microfluidic chip allows us to point out this difference and dissuade any colleague who may be tempted to measure the bubble size of a microfluidic-made foam in the chip, especially in the field of foam templating. Secondly, in polydisperse bubbling to be introduced in the following paragraph, the bubble size quickly varies with the gas pressure p_{gas} preventing us from determining the bubble size at the pressure at which it was formed outside of the chip. Polydisperse bubbling contains us to measure the bubble size inside the chip. Measuring the bubble size inside the chip for monodisperse bubbling provides thus an interesting tool for comparing monodisperse and polydisperse bubblings.

Microfluidic bubbling for polydisperse foams Using the same microfluidic set-up as presented in Fig. 4.3, we are able to produce foams with specific bubble size distributions by setting periodic gas pressures around a constant value $<p_{gas}>$ according to $p_{gas}(t) = <p_{gas}> + [p_{gas,max} - <p_{gas}>] \sin(2\pi t/\tau)$ mbar. The idea is based on the fact that if one can use microfluidics to obtain monodisperse foams with a specific bubble size at a given gas pressure p_{gas}, varying the gas pressure in a controlled manner should yield foams with controlled polydispersities. As mentioned in the previous paragraph, one needs to look at the bubbles inside the chip when varying the gas pressure in order to quantify the influence of the pressure change on the bubble size. Figure 4.5 (bottom) shows the bubble size—measured in the microfluidic channel— as a function of time t for two different gas pressure variations as plotted in Fig. 4.5 (top). The chosen gas pressures follow sinus functions with different amplitudes so that $p_{gas}(t) = 1250 + 150 \sin(\pi t/2)$ mbar and $p_{gas}(t) = 1350 + 250 \sin(\pi t/4)$ mbar. The average gas pressure $<p_{gas}>$, the gas pressure amplitude $p_{gas,max} - <p_{gas}>$, and the period τ constitute the parameters that one can vary in order to tune the bubble size distribution.

For the sake of clarity, we will distinguish the gas pressure functions according to their average pressures $<p_{gas}>$, which are 1250 mbar and 1350 mbar, respectively. Despite the fact that the average pressures of both pressure functions differ by 100 mbar, the baselines of the bubble size in the microfluidic channel lies around 210 μm for both pressure functions. For comparison, the bubble sizes in the channel for constant pressures of 1250 and 1350 mbar are both close to 250 μm. Moreover, for $<p_{gas}> = 1250$ mbar (filled triangles) the bubble size reaches a maximum of 250 μm and a minimum bubble size of 180 μm. The variation of the bubble size resembles the sinusoidal shape of the variation of the pressure, i.e. an extremum of the pressure is reflected in an extremum of the bubble size. However, for $<p_{gas}> = 1350$ mbar, the bubble size goes up to 290 μm for $p_{gas} = 1600$ mbar, but bubbling stops at 1100 mbar, which is the pressure minimum. Note that although no bubbles are produced at 1100 mbar, periodicity is not per se broken, since bubbling restarts as soon as the gas pressure increases again. Although the minimum pressure is 1100 mbar in both cases, microfluidic bubbling shows drastically different behaviours at 1100 mbar, even when compared to monodisperse bubbling, for which a gas pressure of 1100 mbar results in bubbles with a diameter of ca. 250 μm (see Fig. 4.4). The maximum bubble sizes are 250 and 290 μm at both pressure maxima, i.e. at 1400 and 1600 mbar, respectively. For monodisperse bubbling, i.e. at constant gas pressure p_{gas}, the bubble size measured in the channel at pressures of 1400 mbar and 1600 mbar are equal to ca. 260 μm in both cases, which corresponds to the upper limit of the bubble size observed in the channel (Fig. 4.4). Thus, varying the gas pressure allows us to reach larger bubble sizes than for monodisperse, continuous bubbling.

The observation of different bubble sizes for the same pressure depending on whether the pressure is constant or not as well as the amplitudes and frequencies,

Fig. 4.5 (top) Time dependence of the gas pressure p_{gas} over a single period τ for two different pressure functions. Note that the pressures are the ones set with the microfluidic software and thus do not account for experimental fluctuations. (bottom) Variation of the bubble size d_{bubble} measured in the microfluidic channel over a single period τ for the two different pressure protocols. A bubble size of 0 corresponds to the absence of bubbling. The insets are pictures of microfluidic bubblings from which the bubble sizes were measured. All scale bars are 500 μm. Figure taken from [2]

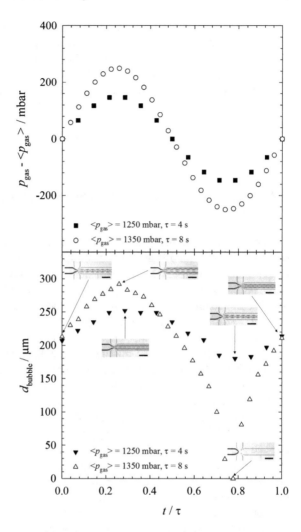

entails the presence of dynamics effects. The speed at which the gas pressure varies affects microfluidic bubbling, i.e. the bubble size distribution. Hence the importance of the period τ, which we have up to now not discussed, on the bubbles size for a given gas pressure function. The period τ is different in the two gas pressure functions and is equal to 4 and 8 s for $<p_{gas}> = 1250$ mbar and $<p_{gas}> = 1350$ mbar, respectively. Practically, one needs to increase the period when increasing the amplitude if one wants to avoid an interruption of bubbling. In order to reach a stable periodic bubbling, one needs to set first the average pressure, a low amplitude (typically 10 mbar), and a long period (typically 10 s). One then increases the amplitude stepwise up until the desired amplitude is reached, while keeping the period constant. Once bubbling is stable for the given amplitude, one may slowly decrease the period. Changing the period too quickly often leads to an interruption of bubbling and requires to start from a constant pressure. One needs, however, to reach the lowest possible period since a too long period obstructs the mixing of the bubbles and thus leads to foams in which small bubbles and large bubbles are separated.

We have seen that the calibration of the chip at constant gas pressures p_{gas} does not help as soon as one applies periodic pressure variations. Thus, for now, the best way to link the polydispersity with the pressure function applied is to directly measure the bubble size distribution of the resulting foams.

4.2 Liquid Foams with Controlled Polydispersities

Characterisation of the Liquid Foams Figure 4.6 shows typical photographs of the liquid foams with different polydispersities generated with the microfluidic set-up presented in Fig. 4.3. Figure 4.6a shows a monodisperse foam with a *PDI* of 3.7% produced at a constant gas pressure of $p_{gas} = 1200$ mbar. Figure 4.6b shows a polydisperse foam with a *PDI* of 14.2% produced with a pressure varying according to $p_{gas}(t) = 1250 + 150 \sin(\pi t/2)$ mbar, and Fig. 4.6c shows a polydisperse foam with a *PDI* of 26.2% produced with a pressure varying according to $p_{gas}(t) = 1350 + 250 \sin(\pi t/4)$ mbar (see Fig. 4.5). Looking at the bubble size distributions of the different foams in Fig. 4.6d, one sees that increasing the amplitude of the pressure variation leads to an increased *PDI*. This behaviour was predicted, although not quantitatively, by the measurements of the bubble sizes presented in Fig. 4.5. The three bubble size distributions are centred around 300 and 400 μm despite the fact that the average gas pressure is different for each bubbling protocol. Thus, with the help of microfluidics, one can produce liquid foams with different *PDI*s but the same average bubble size. However, note that the foam produced with $p_{gas}(t) = 1350 + 250 \sin(\pi t/4)$ mbar has more larger bubbles, which one can trace back to what was already observed in Fig. 4.5, namely the absence of bubbling at the lowest pressures.

Moreover, as the gas pressure p_{gas} decreases, the bubble production rate decreases, which results in foams with fewer small bubbles and more large bubbles. Looking at Fig. 4.6a, one sees that the monodisperse foam is ordered and close-packed whereas both polydisperse foams do not display any order. We already know that monodisperse foams self-order under the action of gravity and confinement [15]. One needs to keep in mind this difference in the order between monodisperse and polydisperse templates as it may affect the mechanical properties of the resulting solid foams. To sum up, microfluidic opens the way to foams with a well-defined bubble size distribution, where one can separately set the average bubble size and the polydispersity. Applied to foam templating one can thus study the influence of polydispersity on the properties of solid foams. We managed to reach a *PDI* of 26.2% with the microfluidic chip used, which is considered as polydisperse but is still lower than the *PDI*s obtained with traditional foaming methods such as mechanical frothing [16]. One may thus aim for larger *PDI*s, which may be reached with microfluidics by increas-

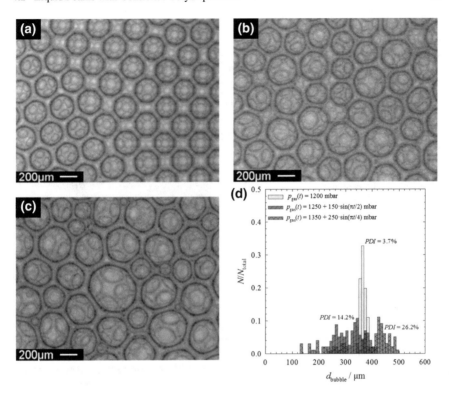

Fig. 4.6 Photographs of **a** a monodisperse liquid foam produced with $p_{gas} = 1200$ mbar, **b** a polydisperse liquid foam produced with an average pressure of $<p_{gas}> = 1250$ mbar and a pressure amplitude of 150 mbar, **c** a polydisperse liquid foam produced with an average pressure of $<p_{gas}> = 1350$ mbar and a pressure amplitude of 300 mbar and **d** their corresponding bubble size distributions. $Q_{chitosan} = 180$ µL min^{-1}. Figure taken from [2]

ing the width of the microfluidic channel or the size of the constriction. We propose as next step in this field to work on the design of chips with geometries allowing for the largest range of bubble size distributions possible.

Cross-linking chitosan foams In the following, we aim to show how the liquid foams produced via microfluidics can be solidified and studied under the spectrum of their polydispersities. We focus on two model systems, namely a monodisperse foam and a poly-disperse foam, which for the sake of clarity, we will simply denote as "monodisperse foam" and "polydisperse foam", respectively. We call monodisperse foam any foam originating from a liquid foam produced with the microfluidic set-up presented in Fig. 4.3 with a constant gas pressure $p_{gas} = 1200$ mbar, an example of which is shown in Fig. 4.6a. We call polydisperse foam any foam originating from a liquid foam produced with the microfluidic set-up presented in Fig. 4.3 with a periodic gas pressure $p_{gas}(t) = 1250 + 150 \sin(\pi t/2)$ mbar, an example of which is shown in Fig. 4.6b.

Fig. 4.7 Gel point measurement via oscillatory rheology at 23 °C for a monodisperse chitosan foam cross-linked with genipin. The gel point is defined as the intersection of the storage and loss moduli, and was determined to be at 47 min. The inset shows the gel point measurement for a bulk chitosan solution cross-linked with genipin. The bulk gel point reads 247 min. Figure taken from [2]

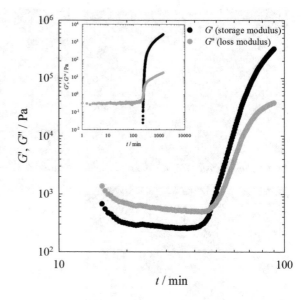

Once we have generated the liquid foam templates, we need to cross-link and freeze-dry them to obtain the desired monodisperse and polydisperse solid foams. Note that the cross-linking step can be left out, and one can directly freeze-dry the liquid foam template, as seen in Sect. 3.4. However, the resulting non-cross-linked foams have weaker mechanical properties; we thus systematically cross-link the liquid foam templates.

We use oscillatory rheology to follow the cross-linking reaction. More precisely, we follow the evolution of the elastic modulus G' and the loss modulus G'' as a function of time as shown in Fig. 4.7. We study both the rheological behaviour of a cross-linking monodisperse chitosan foam and of the chitosan solution itself for comparison (inset). While G' is negligibly small for the bulk chitosan solution before cross-linking, i.e. its value is below the sensitivity limit of the measuring system, the monodisperse foam always has values of G' and G'' which are non-zero, which is a direct consequence of the elastic deformation of the bubbles, but is also observed for viscoelastic liquids in general.

The absence of a storage modulus at low times for the bulk chitosan solution suggests that the storage modulus measured in the foam mainly comes from the contribution of the elastic deformation of the bubbles. The elastic deformation can be approximated by $G' \sim 1/R$ [10]. Both G' and G'' increase steeply during cross-linking with a cross-over which is a measurement for the gel point, i.e. the transition from a liquid-like to a solid-like state. The chitosan foam reaches the gel point after 47 min. The value of G' for the bulk chitosan solution at early times lies several orders of magnitude lower than for the corresponding foam, while, as already mentioned, the value of G'' is negligibly small. This behaviour is expected for a viscous liquid not containing bubbles. As for the foam, G' and G'' increase during cross-linking and one can measure a gel point, which is reached after ca. 4 h (247 min). The bulk gel point is significantly longer than the gel point of the corresponding foam.[2] Surprisingly, the values of both moduli are $\sim 10^2$ Pa higher for the foamed hydrogel than for the bulk hydrogel, which we cannot explain. Since we were interested in the gelation time, we did not wait for the system to fully cross-link in order to look at the values at the plateaus, which would have helped provide a hindsight on that matter.

The Gibbs elastic modulus increases during cross-linking up until a threshold value is reached from which the foam becomes stable: the foam destabilization mechanisms no longer induce foam ageing [8, 18]. One aims for a full conservation of the foam morphology through solidification, that is as little ageing as possible, and thus the earlier the threshold value for the Gibbs elastic modulus is reached, the more the solid foam will resemble its liquid template. However, a too fast gelation is not wished for either, as one needs to avoid that the chitosan solution solidifies in the microfluidic channel, interfering with the generation of bubbles. Since microfluidic bubbling is very pressure sensitive any pressure change in the microfluidic system leads to an unwanted change of the bubble size. As a result, any increase in viscosity of the bubbling solution in the microfluidic set-up makes the control of the bubble size distribution difficult. To ensure the reproducibility of microfluidic bubbling, the rheological behaviour of the bubbling solution has to remain constant within the microfluidic set-up. The solution should thus not solidify within the residence time, i.e. the duration between the time at which the bubble is formed in the chip and the time at which the same bubble leaves the microfluidic set-up. As soon as the bubbles are collected in the Petri dish, they arrange in ordered layers of close-packed bubbles. This reorganization takes typically up to 10 min and should not be stopped by an ill-timed solidification of the chitosan solution. The sum of all the time constraints lead us to estimate an optimal gel point of around 20 min.

The reader may have some concerns about the homogeneity of cross-linking throughout the foam. Let us address this issue. Indeed, the flows in the microfluidic channels are laminar, which does not favour the mixing of the chitosan and genipin solutions (see Sect. 2.4). However, both solutions mix outside the microfluidic channels since the foam is collected in a Petri dish. The liquid fraction is always high enough to allow for the bubbles to move and order, proving that hydrodynamic flows occur within the foam. Moreover, genipin has a long time to diffuse throughout the foam to yield a homogeneous cross-linking since we leave the foam to cross-link for 18 hours, time that needs to be compared with a gel point of 47 min. Moreover, we did not observe any inhomogeneity in colour in the gelled foams, which speaks for a relatively good homogeneity of the cross-linking degree throughout the foam, since, as discussed in Sect. 2.5, the intensity of the blue colour is proportional to the cross-linking degree. Barbetta et al. [5–7] have already tried to reach a perfect homogeneity of cross-linking throughout the foam, but with a different approach.

---

[2]The literature lacks studies accounting for cross-linking of polymers within a foam and the comparison between foamed and bulk hydrogels. We can only hypothesise, thus, that the polymer which is confined within the Plateau borders cross-links faster due to confinement effects. An interesting experiment that could help confirm this theory would be to follow the evolution of. G' and G'' for foams with various liquid fractions φ, all other parameters being kept equal.

Instead of letting the foam cross-link for a long duration, they freeze-dried the foam right after its formation, in order to solidify the template before foam ageing kicks in. The foam is subsequently cross-linked by soaking the foam in a cross-linker solution dissolved in a non-solvent, e.g. a mixture of ethanol: water $= 80 : 20$ for genipin. However, although cross-linking the foam by soaking it in a cross-linker solution addresses the inhomogeneity problem throughout the foam arisen by microfluidic bubbling, it arises an inhomogeneity problem at a lower scale. Indeed, cross-linking the foam in its solid state implies a lack of mobility of the polymer chains, which hinders cross-linking. Thus one expects a higher cross-linking degree at the outer regions of the pore walls compared to the inner regions. To finish this debate, the homogeneity of cross-linking can only be proven by micromechanical analysis such as AFM probing of the cell walls in different regions of the polymer foam in order to verify that the elastic modulus is constant throughout the sample [11]. However, we did not carry out these time-consuming measurements which require an experimental set-up not at our disposal.

Let us go back to the cross-linking procedure followed in the present work. Gelation needs to occur quickly enough to arrest the morphology of the foam template before foam ageing becomes too important. It is thus important to get a grasp of the timescales involved in foam ageing. We assessed the stability of the liquid foam template by looking at the evolution of the relative foam height h/h_0 with time (see Fig. 4.8). We look at both monodisperse and polydisperse foams for comparison, since a different bubble packing may modify the liquid fraction and thus influence foam ageing. The foams were collected in ca. 20 min during which the foam has time to drain. However, for a better comparison, we set the time $t = 0$ min at the time the whole foam is collected. Note that the foams studied here do not contain any C_6F_{14}, not to bias the foam height measurements. Indeed, as discussed in Sect. 3.3, the presence of C_6F_{14} in the bubbles induces a swelling of the upper layers which would bias the measurement of the foam height. Because C_6F_{14} acts against coarsening, and since coarsening does not affect the overall foam height, we decided to run the foam stability tests without C_6F_{14}.

The foams have at least 80% of their initial height after 50 min, even without cross-linking to arrest foam ageing. Within that timeframe, drainage is the principal mechanism for foam ageing, visible through the increasing height of the drained phase. Since drainage affects the liquid fraction and thus the shape of the bubbles, but not their sizes, and knowing that the gel point of the foams is at ca. 50 min, cross-linking is quick enough and fits with the timescales involved in foam ageing. Moreover, the foams do not collapse below 75% of their initial height at longer times, i.e. after 6 h, which confirms a good foam stability despite the absence of mechanical stabilisation via cross-linking. One also sees that the *PDI* does not significantly affect liquid foam stability.

The large bubbles to be seen in Fig. 4.8a, b after 60 min are the result of coarsening and maybe partially coalescence. However, we recall that the experiments presented in Fig. 4.8 were carried out without C_6F_{14}. As a result, the liquid foam templates used to produce solid foams are not submitted to such a strong disproportionation, which we will see by looking at the solid samples.

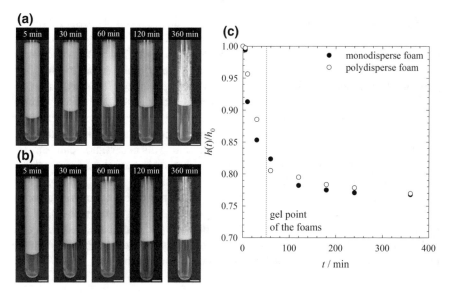

Fig. 4.8 Pictures of **a** monodisperse and **b** polydisperse foam produced via microfluidics at different times after their generation without genipin. The scales bars are 1 cm. **c** Plot of the foam height over the initial foam height as a function of time for the monodisperse and polydisperse foams presented in **a** and **b**. For clarity, not all the pictures used to draw (**c**) are displayed in **a** and **b**. The dotted line in **c** is a guide to the eye and marks the gel point of a monodisperse chitosan foam cross-linked with genipin as determined in Fig. 4.7. Figure taken from [2]

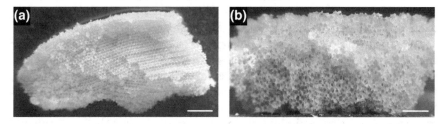

Fig. 4.9 Pictures of **a** monodisperse and **b** polydisperse chitosan foams cross-linked with genipin. The chitosan foams were cross-linked for 18 h at room temperature to produce foamed hydrogels which were frozen in liquid nitrogen to render the sample brittle enough to be broken. The scale bars are 2 mm. Figure taken from [2]

Let us summarise the different timescales involved in foam templating, i.e. the residence time, the gel point and the lifetime of the foam template. One sees that due to a residence time short enough and a gel point long enough the solution does not block the microfluidic device and leaves enough time for the foam to organise. The gel point is however short enough to allow for the foam to solidify before ageing sets in. As a result, the template retains its morphology throughout cross-linking, as one can verify by looking at Fig. 4.9.

Figure. 4.9 shows macroscopic images of monodisperse and polydisperse hydrogel foams. The foams were frozen in liquid nitrogen to solidify the sample and to be able to cut them without damaging the foam structure. Thus, the whiter areas visible on both samples are due to the formation of ice crystals on the frozen hydrogels and should not be linked with the cross-linking density. The samples shown in Fig. 4.9 were randomly chosen and are typical examples of the hydrogel foams before freeze-drying. On the one hand, the monodisperse foam has polyhedral pores which are arranged in ordered layers present over almost the total height of the sample, and one can speak of a crystalline arrangement. This observation confirms the sufficient stability of the foams during cross-linking, an issue that the foam stability tests shown in Fig. 4.8 could not fully address. On the other hand, the polydisperse foam does not show any regularity in the shape of the pores nor any long-range order of the pores. Since the monodisperse and polydisperse foams were cross-linked following the same procedure, i.e. at room temperature for 18 h in a sealed Petri dish and frozen, one can safely conclude that monodispersity is responsible for both the long-range order of the bubbles/pores and their polyhedral shape.

4.3 Monodisperse Versus Polydisperse Solid Foams

Solid chitosan foams Freeze-drying foamed hydrogels such as the ones shown in Fig. 4.9, one obtains then dry solid chitosan foams such as shown in Fig. 4.10a, b. Figure 4.10a shows a monodisperse foam, whereas Fig. 4.10b shows a polydisperse foam. Comparing the foamed hydrogels and the dry foams, one observes a colour change, from a deep blue colour to a greenish blue one. We simply attribute this colour change to the removal of water during freeze-drying. Water, which makes up more than 96% of the mass of the hydrogel foam, is removed during freeze-drying. However, the structure of the resulting cross-linked, freeze-dried monodisperse chitosan foams still corresponds to that of corresponding foamed hydrogel. Indeed, one sees in Fig. 4.10a that the long-range order already observed in the cross-linked foams (see Fig. 4.9a) is not destroyed by freeze-drying. The same observation holds true for polydisperse cross-linked foam.

Figure 4.11 provides a closer look at the porous structures of the monodisperse and polydisperse solid foams via SEM pictures, along with their respective pore size distributions. The relevant parameters of the monodisperse and polydisperse solid foams as well as of their corresponding liquid templates are summarized in Table 4.1. The monodisperse pores are polyhedra with the structure of rhombic dodecahedra, which is sketched in Fig. 4.12b [4]. The crystalline order of the pores observed in Figs. 4.10a and 4.11a is that of an FCC order. Although already reported for monodisperse polyHIPEs [13, 20], such a structure is not usually observed in a solid foam originating from a liquid foam template. Indeed, above a liquid fraction of $\varphi = 0.06323$ the FCC structure is energetically more favourable than the Kelvin structure, which is a BCC structure (see Fig. 2.8 in Sect. 2.1.3). This FCC structure

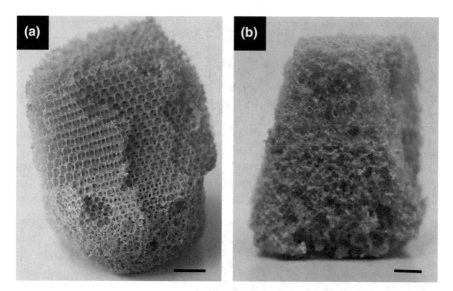

Fig. 4.10 Photographs of **a** monodisperse and **b** polydisperse solid chitosan foams. The scale bars are 1 mm. Figure taken from [2]

is obtained due to the fact that the monodisperse liquid foam template has a liquid fraction of 0.11 which is high enough for the formation of a stable FCC order in the liquid state. Since the template is gelled and frozen before being freeze-dried, the position of the bubbles remains fixed during the drying process. One thus obtains regions with self-ordered bubbles having an FCC order, which, in turn, results in rhombic dodecahedra cells.

In order to calculate the pore volume of the monodisperse foam from a 2D SEM picture, one first needs to measure the centre-to-centre distance between two neighbouring pores d_{cc} as sketched in Fig. 4.12a. All edges of a rhombic dodecahedron are of equal length L, which one can calculate from the centre-to-centre distance d_{cc} as shown in Fig. 4.12b. The pore volume V_{pore} is then given by [4]

$$V_{pore} = \frac{16L^3}{3\sqrt{3}}. \tag{4.1}$$

One can calculate the pore diameter, which is defined as the diameter the pore would have if it were spherical, from the pore volume. We use the pore diameter and not directly the centre-to-centre distance d_{cc} to be as accurate as possible. Indeed, a 2D SEM picture is only the projection of a 3D structure. The centre-to-centre distance d_{cc} can be well measured from a 2D projection, while this is impossible for L as the edges are not in the plane of the cut and not always well defined, due to large pore nodes and a deformation of the pore walls. The calculation of the pore diameter from its volume by considering that the pore is spherical allows us to make a comparison with the bubble size that is more relevant in the scope of foam templating.

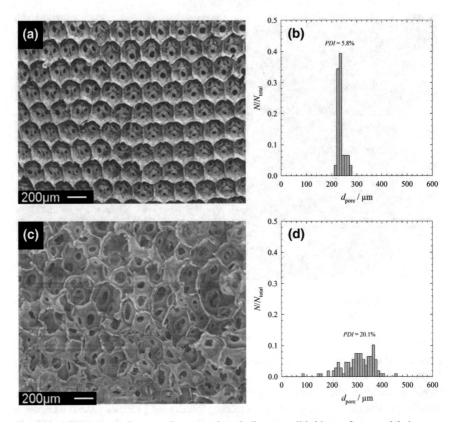

Fig. 4.11 SEM pictures of **a** monodisperse and **c** polydisperse solid chitosan foams and their corresponding pore size distributions, **b** and **d**, respectively. The solid foams result from the solidification of the liquid foam templates shown in Fig. 4.6 a and b, respectively. Figure taken from [2]

The pore size distributions of the monodisperse and polydisperse foams are summarised in Table 4.1. The monodisperse solid foam has a narrow pore size distribution with an average pore size of $<d_{pore}> = 237 \pm 9$ μm and a *PDI* of 5.8%. The polydisperse solid foam has an average pore size of $<d_{pore}> = 304 \pm 61$ μm, with a *PDI* of 20.1%. Since the solid foams are made via foam templating, one has to compare the properties of the solid foams with the properties of the liquid templates they are made from in order to understand the morphological changes that may occur during solidification. Table 4.1 summarises the key parameters of the solid foams along with those of their liquid counterparts. For the monodisperse foam, the liquid template has an average bubble size of $<d_{bubble}> = 364 \pm 14$ μm, while the solid foam has an average pore size $<d_{pore}> = 237 \pm 14$ μm, which corresponds to a size reduction of 35%. For the polydisperse foam, the liquid template has an average bubble size of $<d_{bubble}> = 322 \pm 48$ μm, while the solid foam has an average pore size $<d_{pore}> = 304 \pm 61$ μm, which corresponds to a size reduction of 5.6%.

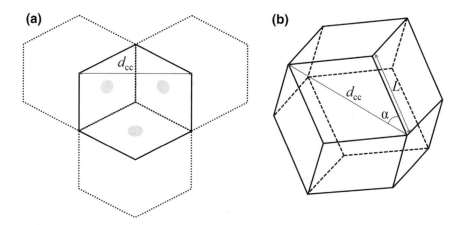

Fig. 4.12 **a** 2D representation of the structure observed in monodisperse solid foams showing four pores. The hexagons drawn with dotted lines correspond to pores belonging to the layer below the observed pore. The grey shapes represent the regions where the interconnects between pores occur. d_{cc} is the centre-to-centre distance between two neighbouring pores and corresponds to the distance separating the two parallel faces oriented upwards of a polyhedral pore. **b** 3D representation of a rhombic dodecahedron. α is the characteristic angle used to calculate L from dcc and is equal to $35.265°$. L is the edge length from which one can measure the volume of the rhombic polyhedron. All edges have the same length L. Figure taken from [2]

Table 4.1 Comparison of liquid and solid foam structures for monodisperse and polydisperse foams. $<d_{bubble}>$ is the average bubble size, PDI is the polydispersity index, φ is the liquid fraction, $<d_{pore}>$ is the average pore size, $<d_{window}>$ is the average window size, ρ is the foam density and ρ^* is the relative density [2]

	Liquid foam			Solid foam				
	$<d_{bubble}>/\mu m$	PDI	φ	$<d_{pore}>/\mu m$	PDI	$<d_{window}>/\mu m$	ρ/g cm^{-3}	ρ^*
Mono	364 ± 14	3.7%	0.11	237 ± 14	5.8%	35 ± 9	0.0113 $\pm$ 0.0017	0.0080 $\pm$ 0.0012
Poly	322 ± 48	14.2%	0.06	304 ± 61	20.1%	77 ± 30	0.0083 $\pm$ 0.0009	0.0058 $\pm$ 0.0006

To explain such a strong difference of shrinkage during solidification between the monodisperse and polydisperse foams, one needs to look at the densities of both systems. Indeed, we measured a strong difference between the densities of the monodisperse and polydisperse foams. The monodisperse solid chitosan foam has a density of $\rho = 0.0113 \pm 0.0017$ g cm^{-3}, while the polydisperse solid foam has a density of $\rho = 0.0083 \pm 0.0009$ g cm^{-3}. We calculate the relative densities of the foams by dividing the densities of the foams by the density of cross-linked chitosan, which is $\rho_c = 1.4207 \pm 0.0031$ g cm^{-3}.

One may correlate the observation that the monodisperse foams are denser than the polydisperse foams to the liquid fractions of the liquid templates. Indeed, the monodisperse liquid template has a liquid fraction $\varphi = 0.11$, while the polydisperse liquid template has a liquid fraction $\varphi = 0.06$. The lower liquid fraction of the polydisperse liquid foam is not surprising in view of the physics of packing: in a polydisperse foam, the smaller bubbles can fill the voids between the bigger ones, which yields a higher packing density. Moreover, considering that the monodisperse foam contains more material than the polydisperse foam, it is not surprising that it shrinks more since there is more material that is subject to shrinkage. Note that the relative densities of both systems are very low, which translate into porosities above 99%. Foam templating usually leads to materials with porosities between 40 and 90% [1]. The high porosity values originate from the sum of (i) the macroporous nature of the chitosan foams and (ii) the low amount of polymer in the continuous phase of the liquid foam template, namely 3.43 wt.% of chitosan. As a result, the struts of the solid foams are themselves porous; however, the pores are too small to be detected by our SEM. Recent micro-tomography measurements performed by Marco Costantini show that the struts are indeed porous, showing a multiscale porosity [14].

Mechanics of solid foams As stated in Sect. 2.2.3, the most straightforward method for characterising the mechanical behaviour of polymer foams is to conduct compression tests. To apply the scaling laws introduced by Gibson and Ashby [17] one needs to know the density and the elastic modulus of the continuous phase. Figure 4.13 shows a mono-disperse solid chitosan foam with the drained phase at the bottom of the sample, which results from the drainage of the liquid template. The inset in Fig. 4.13c shows that the drained phase has a porosity which does not originate from the bubbles of the liquid foam but the from sublimation of the ice crystals during freeze-drying. The freeze-drying of hydrogels to obtain low-density polymer foams is known as ice templating and the community interested in this templating route is growing [12, 19, 22, 27]. The pores in the drained phase are closed and roughly 20 μm large, which is much smaller than the pores resulting from the bubbles in the liquid foam template. However, although the continuous phase of the solid foam in Fig. 4.13b shows some pores (marked by the red arrows in Fig. 4.13b, they are smaller than the ones observed in the drained phase and not homogeneously distributed throughout the continuous phase. Therefore, the continuous phase is so inhomogeneous that one cannot measure a value for its density of elastic modulus that one could apply for the scaling laws described in Sect. 2.2.3.

The monodisperse foam shown in Fig. 4.13 has a structure that strongly differs from the foams shown in Figs. 4.10a and 4.11a. However, all foams have the same composition and were produced in the same way. The irregular structure of the foam shown in Fig. 4.13 is due to the random cut for microscopy and the fact that the blade has not been frozen enough, which may lead to a local shredding of the pore walls during the cut. The question is why the structure in Fig. 4.11a is so regular. We explain this by the fact that the surfaces usually observed during SEM measurement do not result from a cut with a scalpel but are the result of cracks induced by freeze-drying. Indeed, we noticed very early that, even though cracks appear in the foam

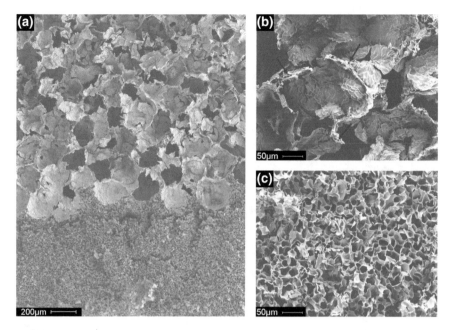

Fig. 4.13 SEM pictures of a monodisperse foam sample **a** showing the solid phase resulting of drainage with insets of **b** the monodisperse foam region and **c** the drained phase, i.e. the freeze-dried chitosan-genipin hydrogel. The red arrows point some pores that originate from the sublimation of ice crystals during freeze-drying

during freeze-drying due to internal stress throughout the foam, these cracks occur in regions of high crystallinity. As a result, the surfaces of these cracks present pores with high monodispersities, long-range orders, and regular shapes.

Comparing the morphology of the drained phase and the morphology of the pore walls and struts, one sees that density and elastic modulus of the drained phase cannot be used for the scalings from Eqs. 2.17 and 2.19. However, for the sake of comparison, we measured the density of the drained phase, which yields $\rho = 0.0600 \pm 0.0006$ g cm^{-3}, and thus $\rho^* = 0.0422 \pm 0.0012$.

Figure 4.14 shows examples of stress-strain curves of a non-foamed freeze-dried chitosan hydrogel, a monodisperse and a polydisperse foam, with a focus on low strains. The scales of the stress axes show that the drained phase reached higher stresses than the chitosan foam samples generated via foam templating. Moreover, the compression curve for the drained phase does not show a straight plateau after the linear region, whereas the monodisperse and polydisperse chitosan foams seem to overlap before densification. The quantitative data that can be extracted from the linear regions of the stress-strain curves, i.e. the elastic modulus E and yield stress σ_y, are summarised in Table 4.2 and graphically shown in Fig. 4.15.

As was already shown in Figs. 4.14 and 4.15a again confirms that the drained phase has a much higher elastic modulus ($E_{\text{drained}} = 6092 \pm 1494$ kPa) than the chitosan

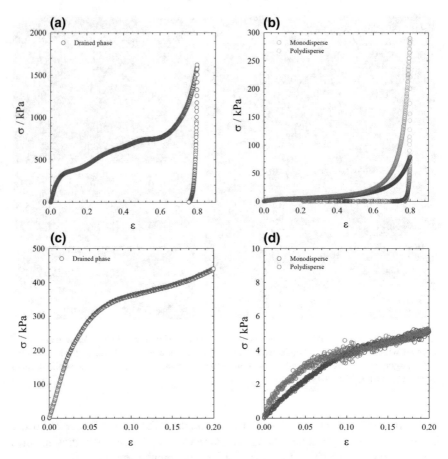

Fig. 4.14 Stress-strain curves of **a** the drained phase region and **b** a monodisperse foam and a polydisperse foam, with foci on the low strains regions of **c** the drained phase and **d** the monodisperse and polydisperse foams. The compressions were carried out at a compression rate of 1 mm/min. We also monitored the stress during decompressions, which were also carried out at a rate of 1 mm/min, without waiting time between compression and decompression

Table 4.2 Relative density ρ^*, elastic moduli E and yield stresses σ_y of the drained phase, poly-disperse and monodisperse foams, along with their respective rescaled values E^* and σ_y^*

	Drained phase	Polydisperse foams	Monodisperse foams
ρ^*	0.0422 ± 0.0012	0.0058 ± 0.0006	0.0080 ± 0.0012
E/kPa	6092 ± 1494	88 ± 22	36 ± 5
σ_y/kPa	423 ± 90	6.3 ± 0.2	3.2 ± 0.2
E^*/MPa	3421 ± 839	49.1 ± 11.9	20.0 ± 2.6
σ_y^*/MPa	238 ± 51	3.54 ± 0.11	1.80 ± 0.45

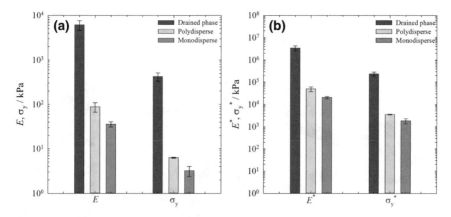

Fig. 4.15 a Elastic modulus E and yield stress σ_y of the drained phase, mono-disperse and poly-disperse foams. The error bars correspond to the standard deviation over three measurements. **b** Rescaled elastic modulus E^* and yield stress σ_y^* of the same samples. The scaling was done by dividing the original values by the squared relative density ρ^{*2} according to Eq. 2.18

foams generated via foam templating ($E_{\text{polydisperse}} = 88 \pm 22$ kPa and $E_{\text{monodisperse}} = 36 \pm 5$ kPa), i.e. the drained phase is much stiffer. Similarly, the drained phase has a much higher yield stress ($\sigma_{y, \text{drained}} = 423.3 \pm 90.3$ kPa) than the chitosan foams generated via foam templating ($\sigma_{y, \text{polydisperse}} = 6.3 \pm 0.2$ kPa and $\sigma_{y, \text{monodisperse}} = 3.2 \pm 0.2$ kPa). This short study of the linear behaviour of the three foams reveals that the drained phase responds to compression very differently from the ways in which the monodisperse and polydisperse foams respond to compression.

However, for any given polymer the density is the main parameter influencing the mechanical response in the linear region (see Sect. 2.2.3). We may apply the scaling law developed by Gibson and Ashby [3, 17] which consists in dividing the elastic modulus E by the squared relative density (Eq. 2.18). Since the yield stress σ_y defines the end of the linear region, one may also apply the same scaling which yields [17]

$$E^* = \frac{E}{\rho^{*2}} \text{ and } \sigma_y^* = \frac{\sigma_y}{r^{*2}}. \tag{4.2}$$

The rescaled values are gathered in Table 4.2 and are graphically shown in Fig. 4.15b. The scaling is used to minimise the effects of density on the linear response to compression, which depends very little on the pore size [17]. Indeed, from the simulations shown in Fig. 2.12, one expects the density to affect the response to compression in the linear region, as opposed to the pore size, the polydispersity or the ordering of the pores. In other words, once the density-based scaling is applied, the stress-strain curves should be close to one another in the linear region. However, one does not obtain similar values of the linear mechanical properties of the different systems. This is not surprising as far as the drained phase is concerned, since the scaling can be used for open-cell foams only while the foam from the drained phase is closed-

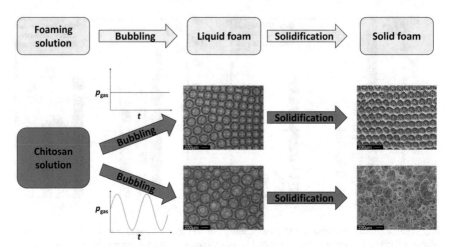

Fig. 4.16 How foam templating can be used to produce solid chitosan foams with a variety of morphologies. In grey is the general foam templating route towards solid foams. In red are the variations from this general route leading to various structures. One can vary the microfluidic bubbling conditions to fine-tune the polydispersity of the liquid foam template, and thus that of the resulting solid chitosan foam. Note that solidification consists in a cross-linking step followed by freeze-drying. Adapted from [2]

cell. An interesting scaling to do in order to gain insight on this matter would be to scale the properties of the monodisperse and polydisperse foams by the properties of their respective continuous phase (Eq. 2.17). However, due to the partial porosity of the pore walls and struts observed in Fig. 4.13b, neither the density of the continuous phase nor its elastic modulus can be constant throughout the material and one thus cannot thoroughly measure them.

The fact that the continuous phase cannot be seen as bulk cross-linked chitosan makes the comparison between the monodisperse and polydisperse foams very difficult. However, the microfluidic method that we developed to produce foams with controlled polydispersities can be applied to various polymer foams, e.g. polymer foams from monomer-based templates, which have an homogeneous continuous phase such as polyurethane foams [24, 25]. A systematic and in-depth study of the mechanical properties of the monodisperse and polydisperse chitosan foams still has to be conducted. Since mechanics is not the focus of this Thesis, we initiated a collaboration with Thierry Roland and Wiebke Drenckhan from the ICS in Strasbourg to go further in the mechanical characterisation.

4.4 Conclusion

We developed in this section a microfluidic bubbling procedure which allows to produce liquid foams with chapter bubble size distributions (monodisperse versus polydisperse) and orderings (ordered versus disordered). We applied the foam templating procedure developed in Sect. 3 to generate solid chitosan foams with controlled pore size distributions and orderings (see Fig. 4.16). We obtained thus two sets of foams with the same chemical composition, comparable average bubble sizes and comparable densities, but different polydispersities and orderings. Interestingly, the monodisperse liquid chitosan foams yielded solid foams with rhombic dodecahedron-shaped pores. Since the differences between the two sets of foams were mainly structural, we could investigate the influence of the polydispersity and ordering on the mechanical properties of the solid foams. Therefore, we studied the foams' mechanical response to compression in the linear region. However, the fact that the values of the elastic moduli are low (below 100 kPa) renders any variation due to the pore size distribution difficult to detect. Thus, even though microfluidic bubbling allows for the generation of solid foams with tailor-made polydispersities, chitosan is not the most adequate material to thoroughly investigate the influence of polydispersity on the mechanics of polymer foams. However, one can apply the developed bubbling method to other systems and search for a system from which one can design solid foams suitable for in-depth mechanical investigations.

References

1. Andrieux S, Quell A, Drenckhan W, Stubenrauch C (2018) Adv Colloid Interface Sci 256:276–290
2. Andrieux S, Drenckhan W, Stubenrauch C (2018) Langmuir 34(4):1581–1590
3. Ashby M (1838) Philos Trans R Soc A 2006(364):15–30
4. Babaee S, Jahromi BH, Ajdari A, Nayeb-Hashemi H, Vaziri A (2012) Acta Mater 60(6):2873–2885
5. Barbetta A, Gumiero A, Pecci R, Bedini R, Dentini M (2009) Biomacromolecules 10(12):3188–3192
6. Barbetta A, Carrino A, Costantini M, Dentini M (2010) Soft Matter 6:5213–5224
7. Barbetta A, Rizzitelli G, Bedini R, Pecci R, Dentini M (2010) Soft Matter 6:1785–1792
8. Bey H, Wintzenrieth F, Ronsin O, Höhler R, Cohen-Addad S (2017) Soft Matter 13(38):6816–6830
9. Calero N, Muñoz J, Ramírez P, Guerrero A (2010) Food Hydrocoll 24(6–7):659–666
10. Cantat I, Cohen-Addad S, Elias F, Graner F, Höhler R, Pitois O, Rouyer F, Saint-Jalmes A (2013) Foams - structure and dynamics. Oxford University Press, Oxford
11. Claesson PM, Dobryden I, Li G, He Y, Huang H, Thorén P-A, Haviland D (2017) Phys Chem Chem Phys
12. Colard CA, Cave RA, Grossiord N, Covington JA, Bon SA (2009) Adv Mater 21(28):2894–2898
13. Costantini M, Colosi C, Mozetic P, Jaroszewicz J, Tosato A, Rainer A, Trombetta M, Święszkowski W, Dentini M, Barbetta A (2016) Mater Sci Eng C 62:668–677
14. Costantini M (2017) Pers Commun
15. Drenckhan W, Langevin D (2010) Curr Opin Colloid Interface Sci 15(5):341–358
16. Drenckhan W, Saint-Jalmes A (2015) Adv Colloid Interface Sci 222:228–259
17. Gibson LJ, Ashby MF (1997) Cellular solids: structure and properties (Cambridge solid state science series). Cambridge University Press, Cambridge
18. Kloek W, van Vliet T, Meinders M (2001), J Colloid Interface Sci 237(2):158–166
19. Martoïa F, Cochereau T, Dumont P, Orgéas L, Terrien M, Belgacem M (2016) Mater Des 104:376–391

20. Quell A, de Bergolis B, Drenckhan W, Stubenrauch C (2016) Macromolecules 49(14):5059–5067
21. Schad T (2017) Surface activity of Chitosan in solution
22. Svagan AJ, Jensen P, Dvinskikh SV, Furó I, Berglund LA (2010) J Mater Chem 20(32):6646–6654
23. Testouri A, Honorez C, Barillec A, Langevin D, Drenckhan W (2010) Macromolecules 43(14):6166–6173
24. Testouri A, Arriaga L, Honorez C, Ranft M, Rodrigues J, van der Net A, Lecchi A, Salonen A, Rio E, Guillermic R-M, Langevin D, Drenckhan W (2012) Colloids Surf A 413:17–24
25. Testouri A, Ranft M, Honorez C, Kaabeche N, Ferbitz J, Freidank D, Drenckhan W (2013) Adv Eng Mater 15(11):1086–1098
26. van der Net A, Blondel L, Saugey A, Drenckhan W (2007) Colloids Surf A 309:159–176
27. Yuan Z-Y, Su B-L (2006) J Mater Chem 16:663–677

Chapter 5
Monodisperse Highly Ordered Nanocomposite Foams

We have seen how to generate monodisperse chitosan foams with interesting morphological properties (Chaps. 3 and 4). However, with an elastic modulus below 100 kPa such foams are mechanically weak in comparison with typical polymer foams. Commercial polystyrene foams have, for example, elastic moduli between ca. 1 and 11 MPa [1]. Wang et al. [11] showed that adding cellulose nanofibres (CNF) to a chitosan-based solution one obtains improved mechanical properties of the resulting solid foams. We thus followed a similar route to improve the mechanical strength of the monodisperse chitosan foams. Indeed, we incorporated different amounts of quaternised CNF into the 4 wt% chitosan solution used in Chap. 4 to produce monodisperse highly ordered nanocomposite foams. The work described in the present chapter follows the work done with negatively charged CNF described in Appendix A.1. This negatively charged CNF showed promising results but had a too low solubility in the chitosan solution and had the tendency to phase separate. Following the advice of Prof. Lars Berglund in KTH Stockholm, we decided to use instead a positively charged CNF which would not present an electrostatic attraction with the polycation chitosan. As defined in Sect. 2.6, a nanocomposite is a material with a matrix reinforced by a filler having at least one dimension in the nanometre range. The matrix in the present section is chitosan and the filler is quaternised cellulose nanofibres (CNF), which are positively charged and are described in detail in [9]. The nanometre-range dimension of the fibers is their diameter, which are on average 1.6–2.1 nm.

5.1 Chitosan/Cellulose Nanofibre Solutions and Microfluidic Bubbling

Rheological behaviour Figure 5.1 shows the viscosity η as a function of the shear rate $\dot{\gamma}$ for the different solutions used throughout this chapter. The composition of each solution and their corresponding denominations are given in Table 5.1. Note that

© Springer Nature Switzerland AG 2019
S. Andrieux, *Monodisperse Highly Ordered and Polydisperse Biobased Solid Foams*,
Springer Theses, https://doi.org/10.1007/978-3-030-27832-8_5

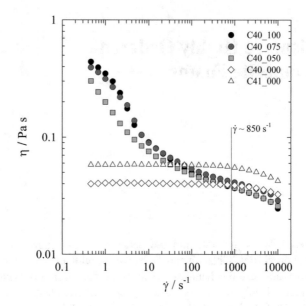

Fig. 5.1 Viscosity η as a function of the shear rate $\dot{\gamma}$ for the different solutions used for the generation of the liquid foam templates. The solutions are C400_100 with $c_{\text{chitosan}} = 4.0$ wt% and $c_{\text{CNF}} = 0.100$ wt%, C40_075 with $c_{\text{chitosan}} = 4.0$ wt% and $c_{\text{CNF}} = 0.075$ wt%, C40_050 with $c_{\text{chitosan}} = 4.0$ wt% and $c_{\text{CNF}} = 0.050$ wt%, C40_000 with $c_{\text{chitosan}} = 4$ wt% and no CNF and C41_000 with $c_{\text{chitosan}} = 4.1$ wt% and no CNF. The straight line shows the shear rate in the T-junction, namely $\dot{\gamma} \sim 850\ \text{s}^{-1}$

Table 5.1 Different solutions used to generate monodisperse foams, with the viscosity at the T-junction η_T and the gas pressure p_{gas} at which the foam templates are produced

Sample	c_{chitosan} / wt%	c_{CNF} / wt%	η_T / mPa s	p_{gas} / mbar
C41_000	4.1	0.000	56 ± 3	1300
C40_100	4.0	0.100	38 ± 3	1700
C40_075	4.0	0.075	42 ± 3	1750
C40_050	4.0	0.050	38 ± 2	1500
C40_000	4.0	0.000	39 ± 3	950

the curve for C40_000 was already shown in Fig. 4.2 (see Chap. 4). One sees that the two solutions without CNF, i.e. C40_000 with $c_{\text{chitosan}} = 4.0$ wt% and C41_000 with $c_{\text{chitosan}} = 4.1$ wt%, present a Newtonian plateau from which one can extract a zero-shear-rate viscosity η_0. One obtains $\eta_0 \sim 0.04$ Pa s for the C40_000 solution and $\eta_0 \sim 0.06$ Pa s for the C41_000 solution. As expected, the solution containing a larger amount of chitosan is more viscous. Moreover, both chitosan solutions display a shear thinning behaviour at high shear rates, i.e. from $\dot{\gamma} > 1000\ \text{s}^{-1}$. Such a shear thinning behaviour is in accordance with previous studies on the rheology of chitosan solutions [2, 3, 10] as well as with our results on the high molecular weight chitosan (see Chap. 3).

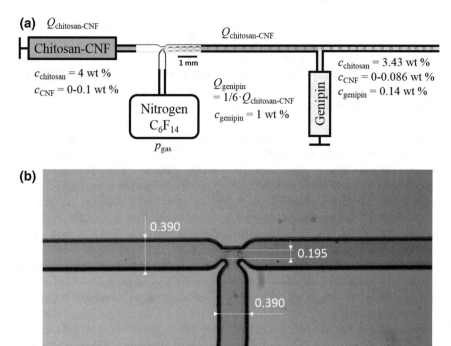

Fig. 5.2   a Generation of monodisperse chitosan/CNF foams using a microfluidic device with a T-junction. **b** Picture of the T-junction used with its relevant dimensions given in mm. The channel depth is constant over the whole chip and equal to 0.190 mm

Looking at the flow curves of the solutions containing CNF, one sees that all the solutions with CNF show a strong shear thinning behaviour over the whole range of shear rates studied. The absence of a plateau at low shear rates is characteristic of yield fluids. For each curve one can distinguish three different regions with three different slopes. The three-region viscosity model has already been observed and described for CNF suspensions [7, 8], but, to the best of our knowledge, it was neither studied in presence of chitosan in particular nor of any other polysaccharide in general. Note that a thorough rheological study requires oscillatory rheometry with the measurement of the storage and loss moduli of the solutions [7]. We chose not to carry out these measurements since such a deep understanding of the interactions between chitosan and CNF is not within the scope of this Thesis. We are, however, interested in the viscosity of the solutions at specific shear rates. We calculated in Appendix A.2 that the shear rate at the constriction of the T-junction used for this work (see next paragraph) is $\dot{\gamma} \sim 850$ s^{-1}. We report the corresponding viscosities at the T-junction η_T in Table 5.1.

Microfluidic bubbling Microfluidic bubbling was carried out in a similar way as in Chap. 4, except for the chip geometry. As shown in Fig. 5.2a, we used a T-junction instead of a cross-flow geometry. The reason is practical: we have never managed to generate a stable flow of monodisperse bubbles with a solution containing CNF using the cross-flow geometry. However, microfluidic bubbling was possible for chitosan solutions containing CNF with a T-junction. The flow curves seen in Fig. 5.1 show that at a shear rate of $\dot{\gamma} \sim 850$ s^{-1}, the chitosan solutions containing CNF have viscosities similar to that of the 4 wt% chitosan solution; a too high viscosity is thus not responsible for the impossibility to produce monodisperse foams with solutions containing CNF. We speculate that the geometry of the chip is responsible. Indeed, we know that the CNFs can be up to 2 µm in length [9] and are elongated due to the high shear rates within the microfluidic channels. While the flow has to follow an elbow in the cross-flow geometry (the direction of the incoming flow is perpendicular the direction of the outcoming bubbly flow), the flow does not change its direction in the T-junction during bubbling (see Fig. 5.2b). Thus, in a T-junction, the cellulose nanofibres may remain elongated in the direction of the flow while in a cross-flow geometry the 90° angle can induce a contraction of the fibres and flow instabilities preventing monodisperse bubbling.

The flow rate of the foaming solution was kept constant at $Q_{chitosan-CNF} = 180$ µL min^{-1} and the flow rate of the genipin solution was $Q_{genipin} = 30$ µL min^{-1}. The gas phase was composed of nitrogen with traces of perfluorohexane C$_6$F$_{14}$ to hinder coarsening.

Figure 5.3a shows the variation of the bubble size d_{bubble} with the gas pressure p_{gas} for the different solutions. The aim of the chip calibration was to determine for each solution the gas pressure to apply to produce a monodisperse liquid foam with the same average bubble size. The bubble size aimed for was ca. 300 µm and was arbitrarily chosen. One sees that, for all solutions, the bubble size d_{bubble} increases with increasing gas pressure p_{gas}. As already observed for the cross-flow geometry (see Figs. 3.6 and 4.4), the bubble size increases steeply at the lowest bubbling pressures (here for the solutions C40_000 and C40_100), before reaching a linear regime at which the bubble size increases slower with increasing gas pressure p_{gas}. One clearly sees that the gas pressures p_{gas} needed to reach bubble sizes between 250 µm and 300 µm are much lower for the solution C40_000 (no CNF, 4 wt% chitosan), than for all other solutions. The difference between the two solutions without CNF (C40_000 and C41_000) can be simply explained by the higher viscosity of the solution containing more chitosan for $\dot{\gamma} \sim 850$ s^{-1}: $\eta_T = 39 \pm 3$ mPa s for C40_000 and $\eta_T = 56 \pm 3$ mPa s for C41_000 (see Table 5.1). For the solutions containing CNF, the gas pressures are also shifted to higher values with increasing CNF concentration c_{CNF}, while their viscosities η_T are all comparable to the viscosity of the C40_000 solution. The viscosity can thus not explain this shift of the gas pressure and one needs to look into the viscoelastic properties of the different solutions to explain this observation. However, a deep investigation of the viscoelasticity of chitosan-CNF solutions was not within the scope of this Thesis as the chip calibration provides sufficient information to produce liquid foam templates with comparable bubble sizes. Figure 5.3b–f show pictures of foam monolayers at the pressures chosen to produce the different liquid foam templates. The gas pressures p_{gas} applied for template production are listed in Table 5.1.

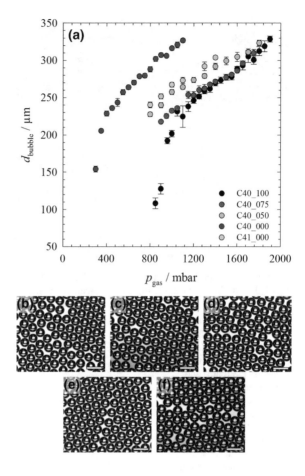

Fig. 5.3 a Bubble diameter d_{bubble} as a function of the gas pressure p_{gas} for the various chitosan solutions. The chitosan flow rate was set to $Q_{chitosan}$ = 180 µL min^{-1}. The error bars correspond to the standard deviations. Pictures of the bubble monolayers at the pressures p_{gas} at which the liquid foam templates are generated, i.e. **b** 1300 mbar for C41_000, **c** 1700 mbar for C40_100, **d** 1750 mbar for C40_075, **e** 1500 mbar for C40_050 and **f** 950 mbar for C40_000. The scales bars are 500 µm. Figure adapted from [5]

5.2 Monodisperse Liquid Chitosan/Cellulose Nanofibres Foams

Once the appropriate parameters for foam production were determined, we characterised the different liquid foam templates. Figure 5.4 shows pictures of a typical monodisperse liquid foam template for each solution. Obviously, the presence of CNF in the chitosan solution does not affect ordering.

Figure 5.5 shows the evolution of the liquid fraction φ of each liquid foam template. One sees that drainage stops and the liquid fraction reaches a plateau for all solutions, but at different times. The values of the plateau liquid fractions are gathered in Table 5.2. The liquid fractions of the solutions without CNF and the solution C40_050 with c_{CNF} = 0.050 wt% decrease sharply with time at short times. However, at long times, the liquid fractions of the solutions without CNF, C40_000 and C41_000 reach the same value of φ = 0.04. For C40_050, the plateau value of the liq-

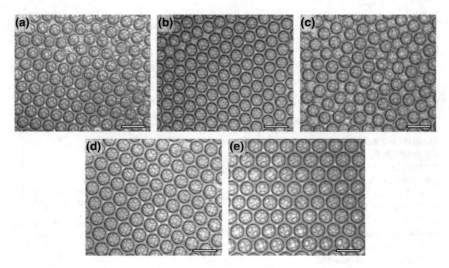

Fig. 5.4 Pictures of the liquid foam templates for each foaming solution: **a** C41_000, **b** C40_100, **c** C40_075, **d** C40_050 and **e** C40_000. The scales bars are 500 μm. Figure adapted from [5]

Fig. 5.5 Liquid fraction φ as a function of time t for the different foam templates. The cross-linker genipin was present in the foams and cross-linking occurred during the experiment. Not to bias the foam volume, we conducted the liquid fraction measurements without perfluorohexane (see Sect. 3.3). Figure adapted from [5]

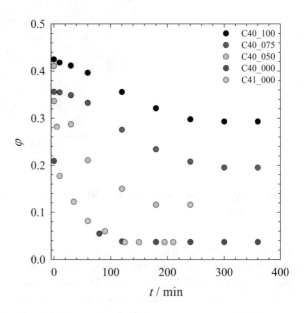

uid fraction is much higher with $\varphi = 0.12$. For the solutions C40_100 and C40_075, containing 0.100 wt% CNF and 0.075 wt% CNF, respectively, one observes two linear regions. The first region is a linear decay of the liquid fraction followed by a plateau once drainage has stopped. The plateau liquid fraction increases with increasing c_{CNF} ($\varphi = 0.20$ for $c_{CNF} = 0.075$ wt% (C40_075) and $\varphi = 0.29$ for $c_{CNF} = 0.100$ wt% (C40_100)).

As previously discussed, the end liquid fraction is not influenced by the viscosity of the liquid. Indeed, the viscosity only affects the time required for the foams to reach the equilibrium liquid fraction profile (see Eq. 2.9 in Sect. 2.1.2). However, looking at the flow curves in Fig. 5.1, we can split the different solutions into two categories: the CNF-free solutions are shear-thinning with Newtonian plateau and a zero-shear-rate viscosity, while the solutions containing CNF are yield fluids. Yield fluids have no zero-shear-rate viscosity and do not flow below a given shear stress called yield stress [6]. One sees that the solutions with a zero-shear-rate viscosity have the same plateau liquid fraction of $\varphi = 0.04$, while the yield fluids have an increasing plateau liquid fraction with increasing CNF concentration. One sees from Fig. 5.1 that the viscosity at low shear rates also increases with increasing CNF concentration, indicating a higher yield stress with increasing CNF concentration [6]: drainage stops at higher liquid fractions for liquids with a higher yield stresses. Another reason for this increase of the liquid fraction with increasing CNF concentration may also be cross-linking. Although we did not measure the gel point for each solution, one can assume that drainage stops upon sufficient cross-linking and the liquid fraction reaches a plateau. The higher viscosities of the solutions with higher CNF concentrations induce a slowing down of drainage. The liquid fraction φ is thus higher when cross-linking arrests drainage. Further investigations to determine the influence of c_{CNF} on the yield stress would bring more insight on this matter. Moreover, oscillatory rheology would allow us to determine whether the gel point is shifted to longer times upon addition of CNF. Although such a shift is expected, because increasing the viscosity results in a decreased mobility of the chitosan molecules within the solution, it would have been interesting to verify and quantify this prediction. We did not carry out these measurements for lack of time. Whichever the reason for the large differences of the liquid fractions, the take-home message of this study is that the CNF concentration c_{CNF} strongly influences the liquid fraction of the liquid template. One thus can expect a strong dependency of the density of the solid foams on the CNF concentration.

5.3 Monodisperse Solid Chitosan/Cellulose Nanofibres Foams

We cross-linked and solidified the liquid foam templates following the same procedure as described in Chap. 4, namely cross-linking at room temperature for 18 h followed by freeze-drying.

Foam density The values of the densities of the solid foams ρ for the different CNF concentration c_{CNF} are gathered in Table 5.2, together with other morphological properties of the solid foams and the corresponding liquid templates. Interestingly, the densities of the solid foams ρ follow the same trend as the liquid fractions φ of their liquid counterparts: the CNF-free foams have comparable liquid fractions in the liquid state and comparable densities in the solid state, and both the liquid

Table 5.2 Comparison of liquid and solid foam properties for foams with different CNF concentrations. $<d_{bubble}>$ is the average bubble size of the liquid foam and φ its liquid fraction. $<d_{pore}>$ is the average pore size of the solid foam and ρ is the foam density. N_{window}/N_{pore} is the average number of windows (or interconnects) per pore in the solid foam. Adapted from [5]

Foam	$<d_{bubble}>$ /μm	φ	$<d_{pore}>$ /μm	ρ / g cm^{-3}	N_{window}/N_{pore}
C41_000	301 ± 6	0.04	282 ± 10	0.012 ± 0.001	3.00
C40_100	310 ± 4	0.29	343 ± 21	0.021 ± 0.002	1.68
C40_075	299 ± 4	0.20	268 ± 28	0.019 ± 0.001	1.98
C40_050	306 ± 3	0.12	309 ± 8	0.015 ± 0.001	3.00
C40_000	303 ± 4	0.04	251 ± 25	0.014 ± 0.002	2.87

fractions and densities increase upon addition of CNF. This illustrates the relations between the liquid and solid foams inherent to foam templating: the flow properties of the liquid solution, which are highly dependent on the CNF concentration, affect the liquid fraction of the liquid template and, in turn, the density of the solid foam.

Foam morphology Figure 5.6a–e show SEM pictures of the resulting solid foams. The pore size distributions measured from the SEM pictures are shown in Fig. 5.6i–v along with the bubble size distributions of the corresponding liquid templates. The samples with $c_{CNF} \leq 0.050$ wt% (C41_000, C40_050 and C40_000) clearly show ordering and retain their monodispersity during solidification, with $PDIs$ of 3.5, 2.5, and 3.3% respectively. Note that the foam with 4 wt% chitosan and no CNF show defaults in its ordering and the sample has been partly shredded (lower right part in Fig. 5.6e), indicating that the cut of the sample was not perfect. This demonstrates once again the limits of SEM for the morphological characterisation of solid foams, which are inherent to the sample preparation.

The foams with 0.075 wt% and 0.100 wt% CNF differ from the other samples due to a broader pore size distribution (see Fig. 5.6 b and c), pores with irregular shapes, thicker pore walls, and a lower number of windows between neighbouring pores (see Fig. 5.6). We quantify the number of windows by defining N_{window}/N_{pore} which is the average number of windows counted per pore. The values of N_{window}/N_{pore} are given in Table 5.2. One sees that the foams up to $c_{CNF} = 0.050$ wt% have in average three windows per pore, which is coherent with rhombic dodecahedra (see Sect. 4.3), but this number decreases with increasing c_{CNF}. One can attribute the decrease of the average number of window per pore for the samples with a larger CNF concentration to the higher densities of the samples (see Table 5.2). Indeed, a higher density implies thicker pore walls, which one can directly observe by looking at the SEM pictures in Fig. 5.6, and renders the breaking of the film less likely. Moreover, looking closer at the shape of the pore windows (see the insets in Fig. 5.6), one sees that the windows of CNF-free foams have smooth oval shapes, whereas the windows in foams containing 0.075 wt% CNF or 0.100 wt% CNF are roughly shaped and frayed. Such irregular shapes do not speak for a slow pore opening mechanism, but rather for of a

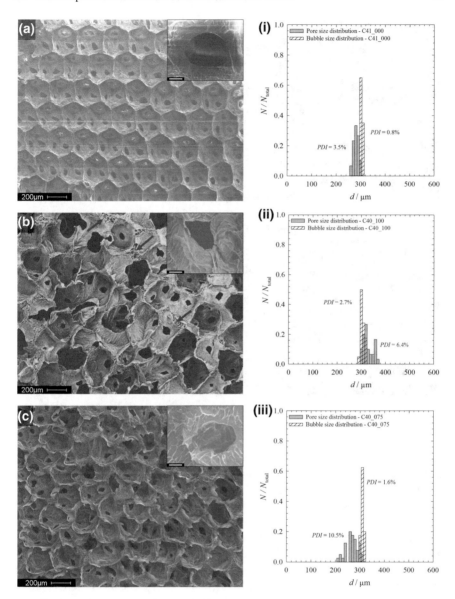

Fig. 5.6 a–e SEM pictures of different solid foam samples, and **i–v** the pore size distributions and bubble size distributions of the corresponding templates generated from the following foaming solutions: **a**, (i) C41_000, **b**, (ii) C40_100, **c**, (iii) C40_075, **d**, (iv) C40_050 and **e**, (v) C40_000. The scale bars in the insets are 20 μm. Figure adapted from [5]

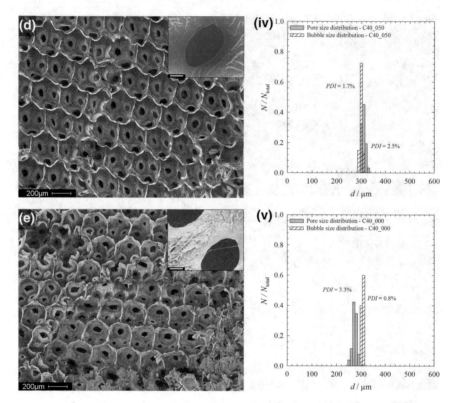

Fig. 5.6 (continued)

sudden breakage of the film.[1] We propose here two possible mechanisms to explain film breakage in the foams with the highest CNF concentrations: (i) The rupture of the film does not occur during solidification due to the enhanced mechanical strength brought about by the addition of CNF (see Appendix A.1), which yields closed-cell foams. The windows observed in Fig. 5.6b and c are thus formed due to the high vacuum applied in the sample chamber during scanning electron microscopy [4]. The foams are thus per se closed-cell, and the opening of the pores is a consequence of the observation method. Micro-tomography, which is carried out in atmospheric conditions, would allow us to verify this hypothesis. (ii) The cellulose nanofibres create internal stresses within the film that induce a folding back of the material once the film breaks. However, the mechanism of film rupture in absence of CNF being itself not well understood, it would be presumptuous to push the discussion further. Indeed, as already discussed in Sect. 2.3.3, the literature lacks convincing explanations for a general film rupture mechanism during the solidification of liquid foams.

---

[1]The SEM picture of the C40_000 foam (Fig. 5.6e) admittedly shows such roughly shaped pores, but in a lesser proportion. We have also mentioned defaults in pore ordering, which can result in thicker pore walls and as a result influence the opening of the pores.

Pore connectivity is an interesting topic that we have not thoroughly investigated and which remains an important gap to fill within our community.

Another peculiarity of the foams containing large amounts of CNF is the porosity of the struts (see e.g. sample C40_100 with $c_{CNF} = 0.100$ wt%). One can see struts which are themselves porous, some of which are pointed out by the red arrows in Fig. 5.6b. We attribute this additional porosity to the higher liquid fraction of the corresponding foam. Indeed, a foam with a high liquid fraction has larger Plateau borders. When freezing such a foam, ice crystals form within the Plateau borders–which become struts in the solid state–and leave pores when the ice crystals sublimate during freeze-drying (see Sect. 2.6).

5.4 Conclusion

The aim of this study was to generate monodisperse solid chitosan foams with improved mechanical properties. We showed how to adapt the foaming process to chitosan solutions containing cellulose nanofibres. The addition of CNF strongly affected the liquid fraction of the liquid templates and, in turn, the density and morphology of the solid foams (see Fig. 5.7). For the sake of comparison, we also investigated a CNF-free chitosan foam with 4.1 wt% chitosan, which has the same solid content as the solution with 4 wt% chitosan and $c_{CNF} = 0.100$ wt%. This 4.1 wt% chitosan foam will serve as control for future mechanical tests. Indeed, adding CNF to the base chitosan solution increases the overall solid content of the foaming solution. We thus have to make sure that any change of the mechanical properties of the solid foams is caused by CNF and not by a simple increase of the solid content.

The mechanical characterisation of the solid foams presented in this chapter is the object of an ongoing collaboration with Prof. Lars Berglund and Lilian Medina at KTH, Stockholm. Early results from our collaboration partner Lilian Medina (see Fig. 5.8) show that the introduction of CNF increases the elastic modulus of the resulting solid foams from an elastic modulus of ca. 120 kPa to ca. 170 kPa (increase of ca. 40%). However, the elastic modulus is not affected by the CNF concentration in the concentration range studied. We observed a different behaviour when adding negatively-charged CNF to the chitosan foams (see Appendix A.1): the elastic modulus increased with increasing cellulose concentration up to an optimum cellulose concentration. The elastic modulus then decreased upon further addition of cellulose. However, bear in mind that the CNF concentrations used in both studies were very different. Such findings are interesting because one sees that CNF does strengthen the chitosan foams and that a low CNF concentration suffices to improve the mechanical strength of the material. However, these results need to be reproduced and the distinction between the foams not containing any CNF, i.e. C40_000 and C41_000, still has to be made.

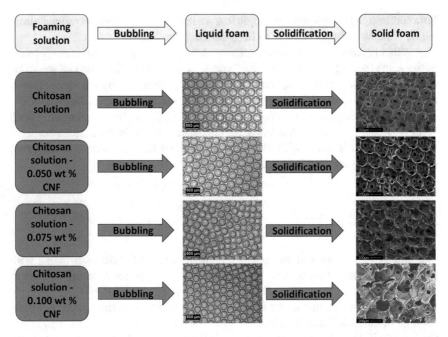

Fig. 5.7 How foam templating can be used to produce monodisperse solid foams with various morphologies. In grey is the general foam templating route towards solid foams. In green are the routes leading to various structures. Note that solidification consists in a cross-linking step followed by freeze-drying. One can vary the pore morphology by adjusting the CNF content in the chitosan solution

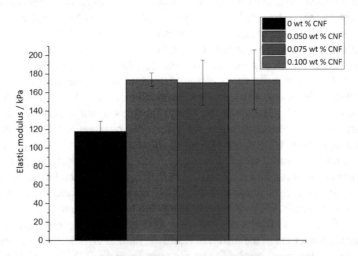

Fig. 5.8 Elastic modulus of chitosan solid foams as a function of the CNF concentration. The samples not containing CNF (C40_000 and C41_000) were not differenciated and the black bar is the average of the measurements for the C40_000 and C41_000 samples. The measurements were carried out at KTH Stockholm by Lilian Medina

References

1. BASF A (2001) Kollicoat IR
2. Calero N, Muñoz J, Ramírez P, Guerrero A (2010) Food Hydrocoll 24(6–7):659–666
3. Cho J, Heuzey M-C, Bégin A, Carreau PJ (2006) J Food Eng 74(4):500–515
4. Drenckhan W (2017) Personal communication
5. Herbst M (2018) Herstellung und morphologische Charakterisierung von celluloseverstä4rkten monodispersen Chitosanschäumen
6. Irgens F (2014) Rheology and non-newtonian fluids. Springer International Publishing, Cham
7. Jia X, Chen Y, Shi C, Ye Y, Abid M, Jabbar S, Wang P, Zeng X, Wu T (2014) Food Hydrocoll 39:27–33
8. Karppinen A, Saarinen T, Salmela J, Laukkanen A, Nuopponen M, Seppälä J (2012) Cellulose 19(6):1807–1819
9. Pei A, Butchosa N, Berglund LA, Zhou Q (2013) Soft Matter 9:2047–2055
10. Testouri A, Honorez C, Barillec A, Langevin D, Drenckhan W (2010) Macromolecules 43(14):6166–6173
11. Wang Y, Uetani K, Liu S, Zhang X, Wang Y, Lu P, Wei T, Fan Z, Shen J, Yu H, Li S, Zhang Q, Li Q, Fan J, Yang N, Wang Q, Liu Y, Cao J, Li J, Chen W (2016) ChemNanoMat, n/a–n/a

Chapter 6
General Conclusions and Outlook

6.1 General Conclusions

Monodisperse Highly Ordered and Polydisperse Biobased Solid Foams: this project title entails various challenges and contraints. To tackle that of biobased solid foams, we only used biobased materials (except for perfluorohexane and acetic acid) for the foaming solution. For the generation of the monodisperse and polydisperse foams, we used foam templating coupled with microfluidics. The high level of ordering was easily reached once we produced monodisperse liquid foams, as monodisperse bubbles spontaneously order. As shown in Fig. 6.1, foam templating allows tailoring the properties of the solid foams by tailoring the properties of the liquid foam templates, which requires continuous feedback loops between the liquid and the solid foams. Indeed, although one can follow a general templating route towards polymer foams, one can modify any step of the process to modify the properties of the final polymer foams, as schematised in Fig. 6.1.

We first had to build up the microfluidic set-up and master microfluidic bubbling to reproducibly produce monodisperse liquid foams. In Chap. 3, we describe how we generated monodisperse liquid chitosan foam templates with bubble sizes from ca. 200 μm to 800 μm. We observed how changing the solidification process affected the morphology of the solid chitosan foams (blue paths in Fig. 6.1). We thus managed to generate mono-disperse solid foams with different morphologies from the same liquid foam template by only varying the solidification procedure. This preliminary work was fundamental to work out the optimal foam templating procedure, but the high molecular weight chitosan used had many drawbacks. For later experiments, we thus used a low molecular weight chitosan which was purer and could be dissolved in higher concentrations instead.

Once we had a satisfying foam templating procedure with the optimal solidification conditions and had reached the first goal, namely generating monodisperse biobased solid foams, we sought to generate polydisperse biobased solid foams for comparison. The aim of this comparison was to find out if monodispersity affects the properties of the foams in general, and the mechanical properties in particular. We

© Springer Nature Switzerland AG 2019

S. Andrieux, *Monodisperse Highly Ordered and Polydisperse Biobased Solid Foams*, Springer Theses, https://doi.org/10.1007/978-3-030-27832-8_6

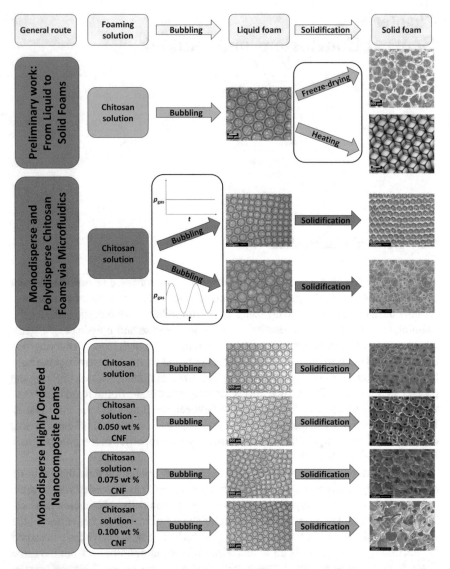

Fig. 6.1 General concept of foam templating (grey) and how we modified different stages of the process (in the black boxes) in each chapter (blue: Chap. 3, red: Chap. 4, green: Chap. 5) to modify the solid chitosan foams

thus produced polydisperse foams with controlled polydispersities via microfluidics by periodically varying the gas pressure. Since microfluidics does not allow to predict the final polydispersity for a given amplitude of the gas pressure, we had to measure the bubble size distribution of the resulting liquid foams and adjust the bubbling procedure accordingly (red paths in Fig. 6.1). Once we managed to produced liquid foam templates with different polydispersities, we solidified them following the same solidification procedure as for the monodisperse foams. We carried out compression tests on the monodisperse and polydisperse solid foams, but could not notice any significant differences between the two systems. Indeed, the elastic moduli of the solid chitosan foams being low, i.e. below 100 kPa, any change of the mechanical response due to the polydispersity should be low and thus difficult to measure; a fortiori considering that the load cell used was 1 kN. A stronger material should thus be used to conduct such a comparative study.

Finally, after having delt with the issue of generating monodisperse highly ordered biobased polymer foams and having developed a procedure to generate biobased polymer foams with controlled polydispersities, we sought to improve the mechanical strength of the said polymer foams. For this purpose, we added cellulose nanofibers to the chitosan solution in order to form monodisperse composite foams. However, changing the composition of the foaming solution required to adjust the microfluidic bubbling conditions to the flow properties of the new solution. We found that the cellulose content strongly affected the liquid fraction of the liquid template, and, in trun, the density and morphology of the resulting solid foam. Early results from collaboration partners confirmed that the addition of cellulose nanofibers strengthens the chitosan foams, but, quite counter-intuitively, to the same extent for all concentrations studied.

6.2   Outlook

Characterisation of the solid foams We have discussed the limitations of scanning electron microscopy for the morphological characterisation of solid foams, namely the fact that one can observe only one section at a time. Micro-computed tomography, or µCT, allows for a three-dimensional scanning of the polymer foam from which one obtains the pore size distribution, the size distribution of the openings and the pore wall thickness. Figure 6.2 shows pictures of the pore size analysis and a 3D reconstruction of a mono-disperse solid chitosan foam such as presented in Sect. 4.3. The measurements were kindly performed by Marco Costantini in Warsaw. With µCT one can visualise the sample in three dimensions, and with computer analysis extract all the morphological data needed. We thus initiated an ongoing collaboration with Marco Costantini and Andrea Barbetta to investigate further the morphological differences between the monodisperse and polydisperse foams generated via microfluidics described in Chap. 4.

We have also seen that the mechanical investigations of the monodisperse and the polydisperse solid chitosan foams did not reveal any significant influence of the polydispersity on the mechanical properties. What was missing in our experiments was (a) the control over the humidity of the room during the experiment and (b) a load cell light enough to measure with precision the mechanical responses of the solid chitosan foams to compression. Both limits can be dealt with at the Institut Charles Sadron in Strasbourg, where an ongoing collaboration with Thierry Roland

Fig. 6.2 **a** Pore size analysis of a monodisperse solid chitosan foam conducted with μCT. **b** 3D reconstruction of the same sample showing the crystallinity of the pores which have an FCC ordering in the [100] direction (from [3])

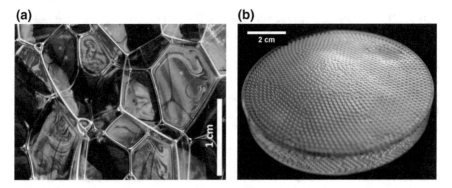

Fig. 6.3 Example of **a** a polydisperse low-density foam and **b** a monodisperse foam generated via microfluidics from the polymer melt DBP. Adapted from [8]

and Wiebke Drenckhan will hopefully provide a better insight on how polydispersity affects the mechanical properties of solid foams.

Foam templating with a polymer melt Thibaut Gaillard showed during his PhD Thesis that the copolymer melt PDMS-*g*-PEG-PPG (commercialised under the name DBP by Gelest Inc.) can form stable free-standing films [8]. This copolymer can also be easily solidified via cross-linking, while remaining transparent [9]. One can also produce bubbles and foams using DBP (see Fig. 6.3), which makes it a good candidate for foam templating. Indeed, we have seen how (i) the weak mechanical properties of chitosan foams and (ii) the porosity of the struts make it difficult to investigate the effects of the polydispersity on the mechanical properties. Using a polymer melt as the continuous phase of a liquid foam template would lead to solid foams with both stronger mechanical properties (as there is more material in the continuous phase) and no porosity in the struts (as no matter is taken away from the liquid foam during solidification). Applying the polydisperse bubbling method developed in Chap. 4 to this system may thus help bring the answers we could not find by investigating chitosan foams.

Weaire-Phelan solid foams Still focusing on foam mechanics, one may use the DBP polymer melt to explore structures never obtained. A good example is the Weaire-Phelan structure, already discussed in Sect. 2.1.3 (see Figure 2.7). Although Gabrielli et al. [7] managed to generate a liquid foam having this structure, there is, to the best of our knowledge, no experimental example of a solid foam having the Weaire-Phelan structure. The procedure for the generation of liquid foams with the Weaire-Phelan structure is well explained in [7]. Thus, one should be able to follow the same procedure to generate DBP liquid foams with the Weaire-Phelan structure and solidify them. It would then be interesting to experimentally investigate the mechanical properties of solid foams which have the Weaire-Phelan structure.

Applications of monodisperse solid foams One could also look into the possible applications of the methods developed in the Thesis at hand. An interesting approach would be to focus on tissue engineering applications. Although chitosan has already been tested at labscale for such applications [2, 5, 6, 11–13], other polymers which are well known for their ability to allow cell growth can be used, such as poly (HEMA), the polymer of 2-hydroxyethyl methacrylate [10]. Indeed the control brought by microfluidic foaming over the bubble size would be of great importance for tissue engineering, and being able to adapt the polymer to the application would be of great interest for the field of biomedicine. Indeed, cell colonisation calls for homogeneous pores with a definite pore size as the optimal pore size differs for each type of cell [1]. A foam serving as scaffold for tissue engineering should also be open-cell to allow for the transportation of nutrients and oxygen and the evacuation of the cell waste. Costantini et al. [4] compared the cell-seeding efficiency of monodisperse and polydisperse scaffolds and showed that the infiltration of the cells is better for a monodisperse foam as compared to a polydisperse foam. However, the authors did not control the polydispersity and the monodisperse and polydisperse foams did not have comparable densities. The polydisperse microfluidic bubbling method developed in Chap. 4 can thus provide monodisperse and polydisperse foams with not only controlled pore size distributions but also comparable densities to be tested for tissue engineering applications.

References

1. Barbetta A, Gumiero A, Pecci R, Bedini R, Dentini M (2009) Biomacromolecules 10(12):3188–3192
2. Christopher GF, Anna SL (2007) J Phys D 40(19):R319
3. Costantini M (2017) Personal communication
4. Costantini M, Colosi C, Mozetic P, Jaroszewicz J, Tosato A, Rainer A, Trombetta M, Święszkowski W, Dentini M, Barbetta A (2016) Mater Sci Eng C 62:668–677
5. Croisier F, Jéróme C (2013) Eur Polym J 49(4):780–792
6. David D, Silverstein MS (2009) J Polym Sci A 47(21):5806–5814
7. Gabbrielli R, Meagher AJ, Weaire D, Brakke KA, Hutzler S (2012) Philos Mag Lett 92(1):1–6
8. Gaillard T (2016) Ecoulements confinés à haut et bas Reynolds: génération millifluidique de mousse et drainage de films minces de copolymères, Paris Saclay, Ph.D. Thesis
9. Gaillard T (2017) Personal communication
10. Kulygin O, Silverstein MS (2007) Soft Matter 3:1525–1529
11. Madihally SV, Matthew HW (1999) Biomaterials 20(12):1133–1142
12. Moglia RS, Holm JL, Sears NA, Wilson CJ, Harrison DM, Cosgriff-Hernandez E (2011) Biomacromolecules 12(10):3621–3628
13. Robinson JL, Moglia RS, Stuebben MC, McEnery MA, Cosgriff-Hernandez E (2014) Tissue Eng Part A 20(5–6):1103–1112

Chapter 7
Experimental

7.1 Chemicals and Preparation of the Solutions

High molecular weight chitosan The high molecular weight chitosan used in Chap. 3 was purchased from Sigma Aldrich with $M_w \sim 300\,000$ g mol^{-1} and has a deacetylation degree DD of $\sim 80\%$ (data from the supplier). The polymer is a light-brown powder and can be dissolved in a 0.05 mol L^{-1} sodium acetate and a 0.2 mol L^{-1} acetic acid solution with a magnetic stirrer to reach a concentration of 1.5 wt% [1]. We filtrated the chitosan solution under vacuum using filter paper (#113 from Whatman, with a pore size of 30 μm) to remove the undissolved particles. This filtration step implies that the final polymer concentration was lower than 1.5 wt%. However, for the sake of discussion and since the chitosan concentration was not a parameter we chose to vary, we will consider that the chitosan concentration remained 1.5 wt%. The surfactant was then added to the solution ($c_{surfactant} = 0.1$ wt%). We then dissolved 0.2 wt% genipin in the chitosan solution while leaving the solution in an ice bath to prevent gelation. The solution was placed in an ultrasound bath (SONOREX SK 100H from Bandelin) for the duration necessary (from 1 to 5 min) to remove the bubbles which had formed during stirring.

Low molecular weight chitosan The low molecular weight chitosan used in Chaps. 4 and 5 was purchased from Glentham Life Sciences Ltd. Its molecular weight is 30 000 g mol^{-1} and its deacetylation degree DD is 90.56% (data from the supplier). The white-light yellow powder was used as received. For the CNF-free chitosan solutions, the chitosan was dissolved in a 1 vol% acetic acid solution with a magnetic stirrer for at least 2 h. To ensure a good dissolution, we treated the chitosan solution with an ultrasonic homogeniser SONOPLUS HD2200 from Bandelin for 5 min at a power of 40%. To prevent the solution from heating up, it was placed in an ice bath during homogenisation. The surfactant was then added to the solution ($c_{surfactant} = 0.1$ wt%), and the solution was left in an ultrasound bath (SONOREX SK 100H from Bandelin) to remove the bubbles that formed while stirring the solution.

© Springer Nature Switzerland AG 2019
S. Andrieux, *Monodisperse Highly Ordered and Polydisperse Biobased Solid Foams*,
Springer Theses, https://doi.org/10.1007/978-3-030-27832-8_7

Quaternised cellulose nanofibres The quaternised cellulose nanofibres (CNF) used in Chap. 5 were sent by Lilian Medina from KTH in Stockholm, as a gel-like suspension in water. The cellulose content was 0.14 wt%. To prepare the CNF/chitosan solutions, we first added water and acetic acid to the CNF dispersion to obtain a solution with the desired concentration of CNF in 1 vol% AcOH. The CNF solution was stirred with a magnetic stirrer for 10 min and treated with an ultrasonic homogeniser SONOPLUS HD2200 from Bandelin for 5 min at a power of 40% to reduce the viscosity of the solution. We added the chitosan powder in five steps up to a chitosan concentration of 4 wt%. After each addition of chitosan, the solution was stirred and homogenised with the SONOPLUS HD2200 for 1 min at a power of 40%. The surfactant was then added to reach a concentration of 0.1 wt% with respect to the solvent and the solution was homogenised one last time for 1 min to remove any remaining bubble in the solution.

Other chemicals The acetic acid was 100% pure and purchased from VWR and the water used was demineralised using an ion exchange column. The surfactant (Plantacare 2000 UP, an alkyl polyglycoside) was donated from Cognis Deutschland GmbH & Co (today BASF) and has an active matter content between 51 and 55%. The alkyl chain contains between 8 and 16 carbons and the head group has on average 1.5 glycoside groups. The cross-linker, genipin, was purchased from Challenge Bioproducts Co., Ltd. and had a purity of 98% (determined by HPLC, data from the supplier). For the work described in Chaps. 4 and 5, we dissolved 1 wt% genipin in 1 vol% AcOH in bidestilled water to reach a concentration of 1 wt% with respect to the solvent. Once the genipin was dissolved, we added the surfactant to reach a concentration of 0.1 wt% with respect to the solvent. Perfluorohexane was 98% pure and purchased from Alfa Aesar.

7.2   Rheometry

We conducted the various rheology experiments using a Physica MCR 501 rheometer from Anton Paar. The temperature was controlled with a Peltier system with a temperature accuracy of 0.1 K and limited the evaporation of the solvent by using a solvent trap. We used different measuring geometries for the viscosity measurements and the gel point measurements.

Rotational rheometry The measurements of the viscosity as a function of the shear rate were conducted with a cone-plate geometry. The cone had a diameter of 49.943 mm, an angle of 1.009°, and was truncated at 101 μm. The measuring gap was 101 μm. The measurements were monitored with the software Rheoplus. The measurements were all carried out three times for each solution at 23 °C. The range of shear rates applied was $0.01–10\,000\,s^{-1}$ and the measuring time varied from 100 to 1 s. Both the shear rates and the measuring point duration varied according to a logarithmic ramp and 6 points per decade were measured. The flow curves in the body of this Thesis report values of viscosity only from a shear rate of ca.

Fig. 7.1 Viscosity η as a function of the shear rate $\dot{\gamma}$ for the 4 wt% chitosan solution over the whole range of measured shear rates. The three curves correspond to three independent measurements of the same solution

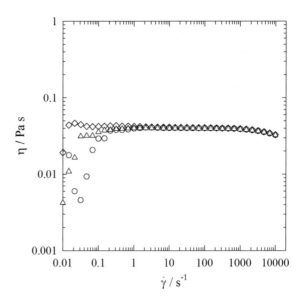

$0.4\,\text{s}^{-1}$ because the viscosity strongly changed between measurements at lower shear rates (see Fig. 7.1). We attribute the low reproducibility at low shear rates to the low sensitivity of the device.

Oscillatory rheometry We used different geometries for the two different chitosans studied in this Thesis. All gel point measurements were carried out at different temperatures and started 5 min after the preparation of the sample. For the high molecular weight chitosan studied in Chap. 3, the composition of the tested solution was 1.5 wt% chitosan cross-linked with 0.2 wt% genipin. The used geometry was a cone-plate, the cone having a diameter of 24.970 mm and an angle of 1.003°. The cone was truncated at 50 µm. The measuring gap was 50 µm and the measurements were performed for a deformation of 1% at a frequency of 1 Hz. For the gel point measurements of the solutions containing the low molecular weight chitosan studied in Chap. 4, we prepared the solution by mixing 3 mL of 4 wt% chitosan and 0.5 mL of 1 wt% genipin to reach the 1/6 volume ratio. We gently shaked the mixture to avoid the formation of bubbles and placed the solution between the two plates of the rheometer. For the gel point measurements of the foam, we simply collected a monodisperse foam from the microfluidic setup and placed it between the plates of the rheometer. The geometry used was plate-plate with a plate having a diameter of 24.975 mm. The measuring gap was 1 mm and the measurements were performed for a 1% deformation at a frequency of 1 Hz.

Fig. 7.2 Surface tension γ as a function of time t of a 4 wt% low molecular weight chitosan solution in 1 vol% acetic acid with 0.06 g L^{-1} Plantacare 2000 UP. The surface tension decreases down to a plateau at $\gamma = 38.2$ mN m^{-1}

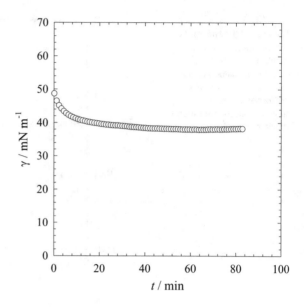

7.3   Surface Tensiometry

The surface tension measurements were carried out using a ST-A1 ring tensiometer from Sinterface. This device is based on the Du Noüy ring method [5]. The measurements were performed at 23 °C. We determined the density of each solution with a DMA 5000 M density meter (Anton Paar) before carrying out the measurements. Figure 7.2 shows a typical example of the evolution of the surface tension with time, which decreases as the surfactant molecules diffuse to the air-liquid interface up until a plateau is reached. The value of the surface tension taken for the studies of the surface tension as a function of the surfactant concentration was the average of the last 10 values of the plateau.

We plotted the surface tensions γ as a function of the surfactant concentration $c_{\text{surfactant}}$ and fitted the data with polynomials of 2nd or 3rd order.

7.4   Microfluidics

Fabrication of the home-made microfluidic chips The self-made microfluidic chips used in Chap. 3 consisted of Cyclic Olefin Copolymers (COC) which has a glass transition temperature (T_g) of 80 °C (COC$_{80}$) and is commercialised under the name TOPAS 8007S-04 by TOPAS Advanced Polymers. COC is a statistic polyethylene-polynorbornene copolymer, which is highly transparent [2, 3] and thus appropriate for microfluidic applications, where live imaging is often necessary [4]. The chips were moulded from a COC$_{170}$ master, which is a COC having a higher T_g of 170 °C,

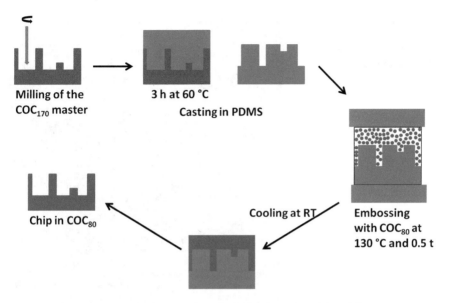

Fig. 7.3 General procedure for the formation of a COC chip. Adapted from [6]

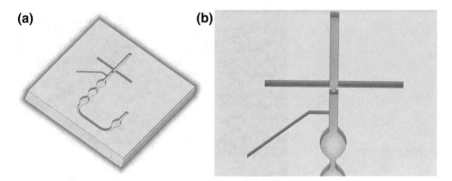

Fig. 7.4 **a** Technical design of the 400 μm COC chip as drawn on SolidWorks. **b** Zoom on the constriction of the same design

allowing the material to be milled. This COC_{170} was also purchased from TOPAS Advanced Polymers and was referenced as TOPAS 6015S-04. The general procedure for the chip fabrication is presented in Fig. 7.3.

First the chips were drawn on SolidWorks with all the required dimensions and were milled on a COC_{170} chip with the help of a milling machine.[1] The technical design of the 400 μm chip is shown in Fig. 7.4.

---

[1] These operations were carried out in the Laboratoire de Physique des Solides, Université Paris-Sud, Orsay, France.

Fig. 7.5 Press equipped
with the temperature control
system for the hot embossing
of the microfluidic chips

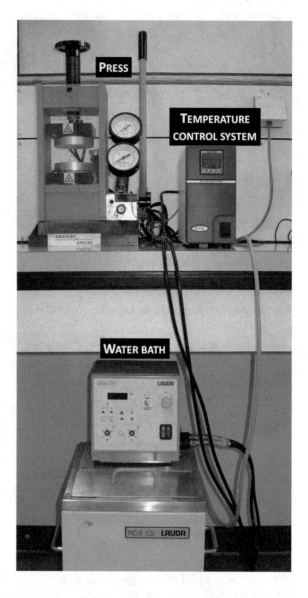

Polydimethylsiloxane (PDMS) male moulds were then formed by placing a silicone oil mixed with a cross-linker (SYLGARD184 Silicone Elastomer Kit, Dow Corning) at a 10:1 weight ratio on the polycarbonate chips. After 3 h in the oven at 60 °C the elastomer had cured and after having been left to cool down at room temperature the PDMS moulds could be extracted. The final COC chips were moulded with COC_{80} from the PDMS master with a manual press equipped with heated platens and a temperature control system (SPECAC, United Kingdom) (Fig. 7.5).

The PDMS casts were inserted into a rectangular aluminium form tailor-made from our workshop, which was filled with COC_{80} granules. The whole was heated to $130\,°C$, without applying any load, for 30 min. A 500 kg load was then applied to the moulds for 20 min while still at $130\,°C$. The load was then released and the moulds were removed from the press to cool down at room temperature.

The chips were then gently separated from the PDMS casts— which could be used again for the next chip— and holes were drilled at the inputs and the output with a 4.0 mm drill bit. Finally the threads in the holes were formed with a screw-tap. The chips were sealed with transparent tape (tesa 64014), and heated to $60\,°C$ for 15 min to promote the adhesion of the films on the chips. The chips were then covered with a polycarbonate plate which was tightened to the chips with pliers in order to ensure that the chips could support the pressure inside the channels. Although this sealing method might seem basic, we could reach pressures up to 1 bar without unsealing the film, which was sufficient enough for the ranges of pressures and flow rates used. Moreover this method made the unsealing of the chip easy enough to allow for a swift and efficient cleaning of the channels, even if the channels were clogged. The chips can thus be used several times and need to be replaced only after 10–20 uses.

Microfluidic bubbling While using the same microfluidic set-up throughout this work, we adapted it for each chapter. Figure 7.6a shows the microfluidic set-up used for the 190 μm glass chip during Chap. 3. Both the gas phase and the liquid phase were pressure controlled, using two separate outlets of the same pressure pump (OB1 Mk2 Pressure Controller from Elveflow) connected to a nitrogen tap. The first outlet of the pressure pump was connected to a glass bottle sealed with a GL45 cap from Vaplock containing perflurohexane (which is a volatile liquid). The bottle was connected to the right inlet of the microfluidic chip. Due to the high vapour pressure of perfluorohexane, traces of perfluorohexane were carried along with the nitrogen flow. The bubbles formed with microfluidics had thus a gas phase composed of nitrogen with traces of perfluorohexane. The chitosan solution containing genipin was kept in a sealed bottle plugged between the chip and the pressure pump.The bottle containing the chitosan solution was kept in an ice bath during bubbling to prevent an early gelation.

For the work presented in Chaps. 4 and 5, we used syringes instead of pressure pumps to push the liquid phases through the channels, as shown in Fig. 7.6b. Chitosan and genipin were dissolved in different solutions so we used one syringe pump for each solution. The syringe pumps (Pump 11 Elite from Harvard Apparatus) allowed us to apply a constant flow rate, i.e. $Q_{chitosan} = 180\,\mu L\,min^{-1}$ for the chitosan solution and $Q_{genipin} = 30\,\mu L\,min^{-1}$. We controlled the bubble size by varying the gas pressure p_{gas}, which we controlled using the software Elveflow Smart Interface. We used Teflon tubings with an outer diameter of 1.6 mm and an inner diameter of 0.5 mm from Techlab.

We monitored microfluidic bubbling by means of a Nikon SMZ- 800 N optical microscope coupled with an Optronis CL600X2 high-speed camera. The software used to record the images was GenICam. The foams were collected in polystyrene

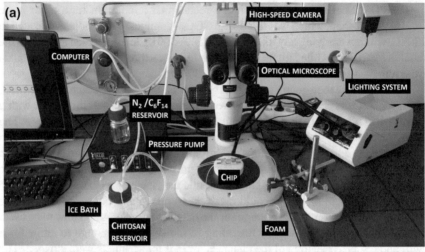

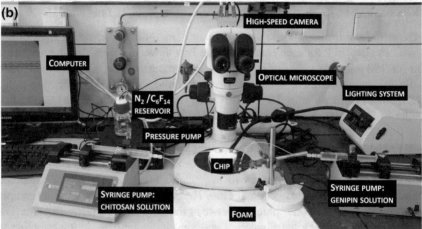

Fig. 7.6 General set-up for the formation of liquid foams via microfluidics, **a** as used for the work described in Chap. 3 with the 190 μm glass chip, and **b** as used for the work described in Chap. 4

Petri dishes with a diameter of 3.5 cm and a height of 1 cm. The time required to collect a foam depended strongly on the liquid pressure and ranged from 1 to 30 min.

Figure 7.7a shows the 190 μm glass chip plugged for cross-flow bubbling. The chip was fixed in a metallic frame (Dolomite), and a connector at each end of the chip tightened the tubings in close contact with the microfluidic chip. If not tight enough, the gas or liquid could get out at the interface between the tubing and the microfluidic channel, which one needed to avoid. The foams were collected outside of the outlet tubing in a Petri-dish, as shown in Fig. 7.7b. One sees that the bubbles form a growing "drop of foam" which falls into the Petri dish once it becomes too heavy. The lifetime of this "drop of foam" varied with the flow rate and the gas pressure, but was roughly

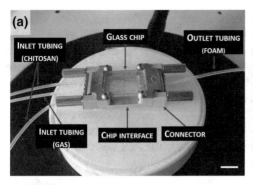
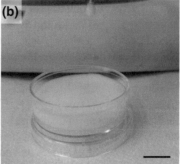

Fig. 7.7 **a** Microfluidic-chip laid under the microscope for the tracking of live bubbling. The inlets are on the left and the outlet is on the right. The chip shown is the 190 μm glass chip with a cross-flow geometry. **b** Output of the microfluidic set-up showing how the foam forms a drop before falling into the Petri dish. The scale bars are 1 cm

between 2 s and 10 s. The fact that the bubbles did not leave the outlet in a continuous manner implies that the pressure at the outlet, and thus within the chip, varied with time. Indeed, the pressure increases during the building up of the "drop of foam" and sinks once it detaches, thus affecting monodispersity. Fortunately, the pressure variations do not decrease monodispersity for the gas pressures p_{gas} at which we carried out microfluidic bubbling – the bubble size was sensitive to these pressure variations at low p_{gas} ($p_{gas} < 100$ mbar), though. Moreover, moving the microfluidic set-up or even just the output tubing affects the bubble size. We thus had to make sure not to move the set-up the chips were calibrated. Consequently, all chip calibrations valid for one specific set-up and coannot be used if one uses tubings of different lengths or change the height of the output. The calibration of the chips is thus the first thing to do once a microfluidic set-up is built, and has to be repeated for every new set-up.

One also sees in Fig. 7.7b that the foams may start to drain before the Petri dish was filled. We thus sometimes had to remove the drained phase with a syringe to obtain a foam high enough to be characterised. We filled the Petri dish up to the top in order to leave as little gas as possible between the foam and the lid, and sealed the Petri dish with Parafilm.

7.5 Liquid Foam Characterisation

Bubble size distribution The bubble size distributions were determined using the image analysis software ImageJ from pictures taken with the Nikon SMZ-800 N optical microscope. Note that the bubble/pores sizes discussed throughout this work are diameters and not radii. For monodisperse closed-packed foams, the bubble size was determined by measuring the distance between the centres of two bubbles in contact, which we call the centre-to-centre distance d_{cc}. The bubble centre was calculated by ImageJ and is the average position of all pixels belonging to the bubble. This allowed for a more precise determination of the bubble size if the contours of the bubble are out of focus. For polydisperse foams, the bubble size was calculated from the area of the bubble measured with ImageJ. In the absence of a close-packing the centre-to-centre distance this area is not equal to twice the radius. Note that the visible black rings (see e.g. Figs. 3.7, 4.6, and 5.4) are not the contours of the liquid bubbles, but are an optical effect [7]. At least 40 bubbles were measured for each sample to determine the polydispersity index (*PDI*). The *PDI* was used to assess the monodispersity of the liquid and solid foams.

Liquid fraction The liquid fractions of the foams were determined by collecting the foams in a 10 mL graduated cylinder for a given duration Δt. The liquid fraction φ of the foam in the cylinder differs from the liquid fraction set by microfluidic bubbling due to drainage. To measure φ, we determined the total volume of liquid V_1 in the cylinder by first summing the flow rates of the syringe pumps ($Q_{total} = Q_{chitosan} + Q_{genipin}$), which yields $V_1 = \Delta t.Q_{total}$. The volume of the drained liquid $V_{1,drain}$ and the foam volume V_{foam} were measured with the graduation of the cylinder. The volume of the liquid contained in the foam $V_{1,foam}$ was the difference between the total volume and the volume of the drained phase, i.e. $V_{1,foam} = V_1 - V_{1,drain}$. The liquid fraction φ was the ratio of the volume of liquid in the foam and the volume of the foam, i.e. $\varphi = V_{1,foam} / V_{foam}$.

Foam stability The liquid foam stability was assessed by collecting the foams in 10 mL test tubes. We took pictures of the foams right after their formation and after different times. We measured the foam height from the pictures using ImageJ and the foam stability was quantified by reporting the evolution of the liquid foam height $h(t)$ with time. To compare the different samples, we used the normalised foam height $h(t)/h_0$, with h_0 being the initial foam height, and plotted it versus time t.

7.6 Liquid Foam to Solid Foam Transition

Cross-linking The cross-linking procedure followed in Chaps. 4 and 5 was the same for all samples. The Petri dishes containing the foams were sealed with Parafilm and left to cross-link at room temperature for 18 h. In Chap. 3, we investigated the possibility to accelerate gelation by heating the foams. We thus either (i) left the foams for 2 h in an oven at 40 °C followed by 18 h at room temperature, or (ii) left the foams for 2 h in an oven at 40 °C followed by 18 h at 60 °C.

Freeze-drying Once cross-linked, the foams were frozen in liquid nitrogen and freeze-dried in an Alpha 1 – 4 LSC freeze-dryer from CHRIST.

7.7 Solid Foam Characterisation

Pore size distribution and morphology We investigated the structure of the solid foams with scanning electron microscopy (SEM) using a CamScan CS 44 microscope. The solid foams were frozen with liquid nitrogen before being cut with a scalpel. Without this freezing step, the foams were not stiff enough to be cut and were shredded or were compressed upon cutting. Once the samples were cut, we glued them on a sample holder and sputtered them with gold to coat the samples with a conductive layer. The voltage applied was 5 kV for enlargements below 1000x and 15 kV for enlargements of 1000x and larger. The software used to record the images was Edax Genesis. The pore size distributions were measured from the SEM images using ImageJ. For polydisperse foams and samples not showing regular shapes, the pore size was the equivalent diameter calculated from the area of the pore using the formula of the area of a disk. The pore sizes of samples having pores with the shape of a rhombic dodecahedron were measured from the centre-to-centre distance, as explained in Sect. 4.3.

Densities The density of the material constituting the solid foams $\rho_{polymer}$ was measured using a Helium porosimeter AccuPycII 1340 from Micrometrics in Vienna. We conducted the density measurements on chitosan solid foams which were ground into a powder. We measured the foam density ρ_{foam} by cutting the foams into regular shapes, measuring the samples volumes, and weighing them. We measured the volume of the samples by taking pictures of the samples and measuring the different diamensions with ImageJ, using the calliper for scale (see Fig. 7.8). Due to the low weight of the samples (of the order of microgramme), the density measurements had a large error. We thus repeated the measurements as many times as possible, i.e. at least 6 times for each foam.

Mechanical properties The stress-strain curves shown in Chap. 4 were measured in Vienna with a universal mechanical tester (Instron 5969) at room temperature. The load cell used was 1 kN, which is admittedly too high for samples as mechanically weak as the chitosan foams studied, but it was the lightest load cell available. The strain was set from 0 to 80% and the compression rate was $1\,\text{mm min}^{-1}$. The measurements were carried out three times for each foam.

The stress-strain curves shown in Appendix A.1 were measured in Stuttgart using an MCR 501 rheometer from Anton Paar mounted with a plate-plate geometry. The measurements were stress-controlled, i.e. we applied a normal force from 0 N to 30 N during 3000 s, with 300 measuring points. We calculated the stress which corresponds to the given normal force using Eq. 2.15 and the strain using Eq. 2.16.

The elastic moduli were calculated by fitting the linear regions of the stress-strain curves, an example of which is shown in Fig. 7.9. One can also fit the plateau region. The intersect of the straight line (red line in Fig. 7.9) with the fit of the linear region is used to determine the yield stress σ_y.

Fig. 7.8 Picture of a foam
sample used for the
determination of the
sample's volume with
ImageJ

Fig. 7.9 Stress-strain curve
of a monodisperse solid
foam including a linear fit of
the linear region (black line).
The slope of this fit yields
the elastic modulus E (here E
= 22.9 kPa). The second fit is
a linear fit of the plateau
region (red line) as defined in
Sect. 2.2.3. The intersection
of both fits is the yield stress
σ_y (here σ_y = 4.2 kPa)

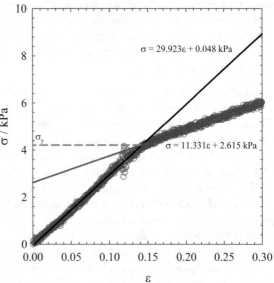

References

1. Calero N, Muñoz J, Ramírez P, Guerrero A (2010) Food Hydrocoll 24(6–7):659–666
2. Khanarian G, Celanese H (2001) Opt Eng 40(6):1024–1029
3. Lamonte RR, McNally D (2001) Adv Mater Process 159(3):33–36
4. Nunes PS, Ohlsson PD, Ordeig O, Kutter JP (2010) Microfluid Nanofluid 9(2–3):145–161
5. Rusanov A, Prokhorov V, Möbius D, Miller R (eds) (1996) Interfacial tensiometry. Elsevier, Amsterdam
6. Testouri T (2012) Highly structures polymer foams from liquid foam templates using millifluidic lab-on-a-chip techniques, Université Paris-Sud XI, Ph.D. Thesis
7. van der Net A, Blondel L, Saugey A, Drenckhan W (2007) Colloids Surf A 309:159–176

Appendix

A.1 Polydisperse Nanocomposite Foams

Before working with the quaternised cellulose nanofibers as described in Chap. 5, we first tested the effects of adding non-modified enzymatic CNF, i.e. negatively charged CNF, to chitosan foams. The enzymatic CNF (abbreviated e-CNF, as opposed to the quaternised CNF, simply abbreviated CNF, reported in Chap. 5) was kindly provided by Lilian Medina, from the KTH in Stockholm, as a gel-like 1.58 wt% dispersion in water. We managed to dissolve up to 0.4 wt% of e-CNF in the chitosan solution. We generated polydisperse foams using a milk foamer at a rotation speed of 1000 rpm for 30 s. The polymer foams were prepared such that their composition equals that of the monodisperse foams generated via microfluidics (see Chaps. 4 and 5). Practically, 9 mL of a 4 wt% chitosan solution (containing a given amount of e-CNF) were mixed with 1.5 mL of a 1 wt% genipin solution in order to reach a volume ratio of 1/6. The e-CNF concentration $c_{e\text{-}CNF}$ is the concentration with respect to the solvent in the chitosan solution, i.e. before its mixing with the genipin solution dilutes both the chitosan and the e-CNF (see Fig. 5.2).

Polydisperse liquid chitosan/CNF foams To study how the addition of e-CNF to chitosan foams affects their properties, we generated polydisperse chitosan-based solid foams via foam templating. The templates contained 4 wt% of chitosan and different amounts of e-CNF, namely 0, 0.1, 0.2, 0.3 and 0.4 wt%. Figure A.1 shows pictures of the liquid foam templates for each e-CNF concentration. The corresponding average bubble sizes $<d_{bubble}>$ and *PDI*s are summarised in Table A.1. All foams have an average bubble size between 330 μm and 360 μm and a *PDI* above 20%. Interestingly, the bubble sizes remain comparable for all solutions, despite the fact that there is a clear increase of the viscosity of the solutions (simply noticeable by gently shaking the solutions) with increasing e-CNF concentration.

Before looking at how the presence of e-CNF affects the mechanical properties of the solid foams, we studied the stability of the e-CNF-loaded liquid chitosan foams. Figure A.2a–e shows pictures of liquid foams 1 h after their formation for

© Springer Nature Switzerland AG 2019

S. Andrieux, *Monodisperse Highly Ordered and Polydisperse Biobased Solid Foams*, Springer Theses, https://doi.org/10.1007/978-3-030-27832-8

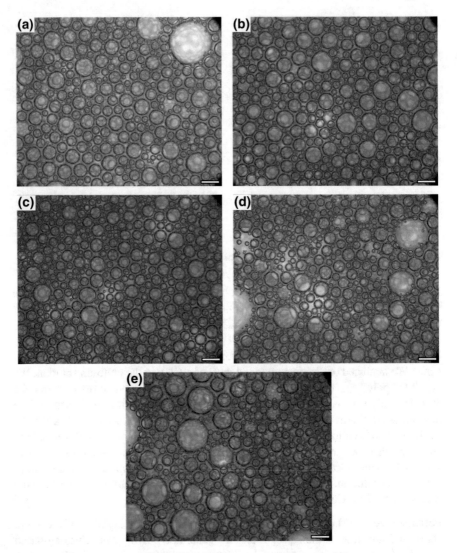

Fig. A.1 Pictures of the polydisperse liquid foam templates with 4 wt% chitosan and different e-CNF concentrations $c_{e\text{-}CNF}$, **a** $c_{e\text{-}CNF} = 0$ wt%, **b** $c_{e\text{-}CNF} = 0.1$ wt%, **c** $c_{e\text{-}CNF} = 0.2$ wt%, **d** $c_{e\text{-}CNF} = 0.3$ wt% and **e** $c_{e\text{-}CNF} = 0.1$ wt%. The scale bars are 500 μm. Adapted from [3]

the different e-CNF concentrations. One observes that below a CNF concentration of $c_{CNF} \leqslant 0.3$ wt% the chitosan/e-CNF solution is turbid, while one sees a clear phase separation of the solution from an e-CNF concentration of $c_{e\text{-}CNF} \geqslant 0.3$ wt%, even at earlier times, i.e. after ca. 10 min. A possible explanation for the phase separation is that the large negatively charged fibres build aggregates with the positively charged chitosan that are large enough to induce a macroscopic phase separation [1]. Note

Table A.1 Average bubble size $<d_{bubble}>$ and *PDI* of the polydisperse liquid foams with different e-CNF concentrations c_{e-CNF}. Adapted from [3]

c_{e-CNF} / wt%	$<d_{bubble}>$ / μm	*PDI* / %
0	356 ± 75	21
0.1	356 ± 83	23
0.2	339 ± 66	20
0.3	332 ± 66	20
0.4	343 ± 75	27

that this phase separation is also observed for the unfoamed solutions, but only after a few days. Foaming the chitosan/e-CNF solutions thus seems to accelerate this phase separation. One may speculate that foaming the solution via mechanical stirring facilitates the formation of chitosan/e-CNF complexes. Another likely reason for this early phase separation may be the confinement of the chitosan and the fibres between the gas bubbles which facilitates the formation of complexes. The relative foam height h_t/h_0 as a function of time t is shown in Fig. A.2f. One sees that, as expected, the relative foam height decreases with time as drainage occurs. However, the higher the e-CNF concentration in solution, the lower is the relative foam height, i.e. the lower is foam stability. Moreover, for the samples showing phase separation, i.e. for $c_{e-CNF} \geqslant 0.2$ wt%, the foams destabilised within 10 min but no longer collapsed from that point on. This sudden destabilisation may be linked with the formation of complexes and phase separation by inducing a local depletion of polymer and fibers which destabilises the foam films. Since the drained phase is slightly turbid (see Fig. A.2b and c) one may also argue that there is also a formation of complexes for 0.1 and 0.2 wt% e-CNF which destabilises the foam. In conclusion, the foams with the best stability are the ones without e-CNF.

Bearing these observations in mind, we now look at the solidification of the e-CNF-loaded chitosan foams which was carried out via cross-linking at room temperature during 18 h and subsequent freeze-drying.

Polydisperse solid chitosan/CNF foams Once freeze-dried, we cut the foams into regular shapes so that their dimensions could be easily measured. Figure A.3 shows the density of the solid foams as a function of the e-CNF concentration c_{e-CNF}. We measured the foam density by weighing the foams which were cut into regular shapes. The large error bars are due to the low weight of the samples (several milligrams while our balance was precise only to 0.1 mg). Moreover, since chitosan and cellulose are strongly hygroscopic, the foams quickly absorb water from the atmosphere, which influences the weight of the sample. Despite these factors, which reduce the precision of the measurements, one observes an increase of the foam density with increasing e-CNF concentration c_{e-CNF}.

Figure A.4 shows SEM pictures of solid chitosan foams without e-CNF and loaded with $c_{e-CNF} = 0.2$ wt% and $c_{e-CNF} = 0.4$ wt%. Due to the relatively low stability of the liquid foams containing e-CNF we cannot compare the pore size distribution

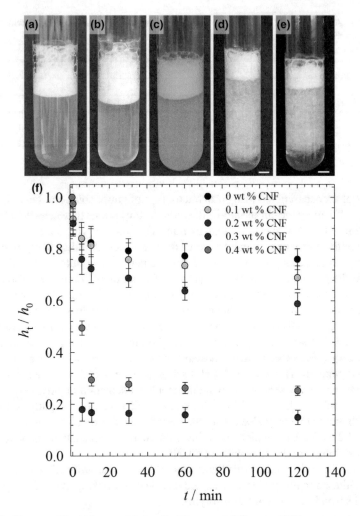

Fig. A.2 Pictures of liquid foams with 4 wt% chitosan and different e-CNF concentrations $c_{\text{e-CNF}}$ 1 h after their formation: **a** $c_{\text{e-CNF}} = 0$ wt %, **b** $c_{\text{e-CNF}} = 0.1$ wt%, **c** $c_{\text{e-CNF}} = 0.2$ wt%, **d** $c_{\text{e-CNF}} = 0.3$ wt % and **e** $c_{\text{e-CNF}} = 0.4$ wt%. The foams did not contain genipin. The scale bars are 5 mm. **f** Relative foam height h_t/h_0 with time t (adapted from [3])

of the solid foams with the bubble size distribution of the liquid foam templates. Indeed, we did not used perfluorohexane during foaming to stabilise the foams against coarsening, which leads to the formation of bubbles as large as one millimetre. One can, however, make an interesting observation by looking at the pore openings of the different samples. Indeed, the foam containing no e-CNF in Fig. A.4a, b has large openings, whereas the chitosan foam containing 0.4 wt% e-CNF in Fig. A.4e, f is fully closed-cell. The foam containing 0.2 wt% e-CNF shows a morphology in between, i.e. with openings but fewer than in the absence of e-CNF. The addition of

Fig. A.3 Density of the solid chitosan foams as a function of the e-CNF concentration $c_{\text{e-CNF}}$ (adapted from [3])

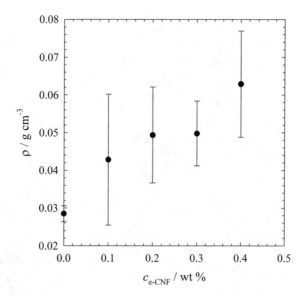

e-CNF seems thus to act against pore opening. One may attribute this phenomenon to the adsorption of e-CNF at the air-liquid interface, which strengthens the thin film enough to prevent its rupture during drying. However, given that the increase in e-CNF concentration leads to a density increase, one may also argue that the closed-pore morphology results in a straightforward manner from the fact that their is more material in the foam as the e-CNF concentration increases. In other words, if the pore walls are thicker, openings are less likely to form upon solidification. The dependency of the density on the e-CNF concentration may originate from a dependency of the liquid fraction on the e-CNF concentration. Unfortunately, we did not measure the liquid fractions of the liquid foams to verify this assumption.

Moreover, looking closer at the surface of the pores (insets of Fig. A.4b, d and f), one notices a morphological modification brought about by the presence of e-CNF. In absence of e-CNF, the surface of the pores is smooth on the micrometre scale, whereas the surface of the pores show bumps and fibrilar structures for the foams containing 0.2 and 0.4 wt% e-CNF. One may attribute the observed bumps and fibrils to the presence of e-CNF at the pore surface, i.e. at the air-liquid interface in the liquid state. We cannot, however, conclude that we directly observe the cellulose nanofibrils since cellulose nanofibrils have diameters below 100 nm (see Sect. 2.6), while the smallest fibrils measurable have a diameter above 500 nm. The fibrilar structures observed may thus result from chitosan-e-CNFs complexes or aggregates of e-CNF yielding complexes large enough to be observed at this scale [2]. The bumps observed in the foams with e-CNFs may originate from a local modification of the rheological properties of the solution due to the cellulose nanofibres and the phase separation already observed in the liquid state (see Fig. A.2) [2]. Both explanations for the bumps on the pore surface and the fibrils require that the e-CNFs have an

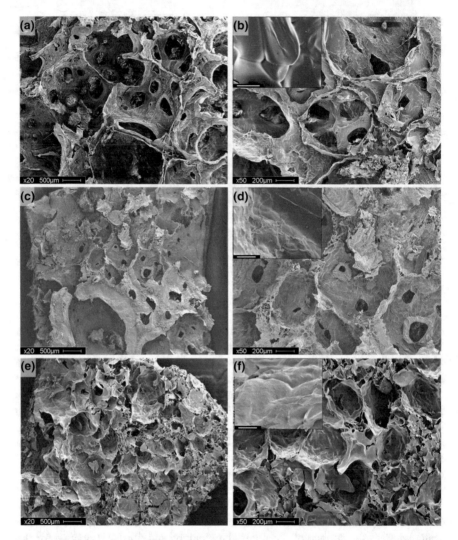

Fig. A.4 SEM pictures of solid chitosan foams with **a, b** $c_{\text{e-CNF}} = 0$ wt%, **c, d** $c_{\text{e-CNF}} = 0.2$ wt% CNF and **e, f** $c_{\text{e-CNF}} = 0.4$ wt% CNF. The scale bars in the insets are 30 μm (adapted from [3])

affinity to the air-liquid interface and coat the surface of the bubbles in the liquid state, supporting our argument for the origin of the closed-cellness of the e-CNF-containing foams. Unfortunately, this remains pure speculation as we did not carry out surface rheology measurements with the enzymatic e-CNF used during the work presented in this section.

Finally, we measured the stress-strain curves for the foams with the different e-CNF concentrations and show an example of each in Fig. A.5a. Figure A.5b shows the variation of the elastic modulus of the foam as a function of the e-CNF concentration

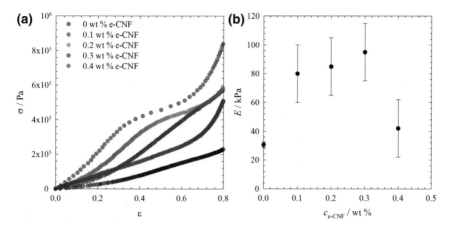

Fig. A.5 **a** Examples of stress-strain curves for solid chitosan foams with different e-CNF concentrations. **b** Elastic modulus of the solid chitosan foams as a function of the e-CNF concentration $c_{e\text{-}CNF}$ calculated from the slopes of stress-strain curves (adapted from [3])

$c_{e\text{-}CNF}$. We attribute the large error bars for $c_{e\text{-}CNF} \geq 0.1$ wt% to the strong hygroscopic character of cellulose (stronger than that of chitosan), as the water absorbed from the atmosphere serves as a plasticiser for the polymer matrix. In other words, the dependency of the mechanical properties on the atmospheric humidity, which we could not control during the experiments, is non-negligible [2]. Despite this experimental impediment, one can extract a trend showing a strong increase of the elastic modulus from no e-CNF to the lowest e-CNF concentration $c_{e\text{-}CNF} = 0.1$ wt%, followed by a slower increase with increasing $c_{e\text{-}CNF}$ until $c_{e\text{-}CNF} = 0.3$ wt%. At $c_{e\text{-}CNF} = 0.4$ wt% the elastic modulus drops down to a value comparable to that of the e-CNF-free foams. This goes against the first expectation that the more e-CNF is present in the composite, the larger is the elastic modulus of the said composite. The lower elastic modulus at 0.4 wt% e-CNF most likely results from a phase separation that left regions within the foam depleted of e-CNF [1]. There is thus an optimal e-CNF content to find in order to get the highest elastic modulus possible, keeping in mind that a small amount of e-CNF suffices to significantly enlarge the elastic modulus.

Adding e-CNF to the chitosan solution during foam templating helps thus increase the elastic modulus of the resulting foams, but also affects the pore morphology. However, using e-CNF has the drawback of forming chitosan-e-CNF complexes that lead to phase separation at high e-CNF concentrations. Note that one needs to avoid phase separation but needs to reach for the best solubility possible of CNF in the chitosan solution to obtain a homogeneous distribution of material in the solid foam. We thus decided not to pursue further with e-CNF and worked instead with quaternised, i.e. positively charged, CNF which was also kindly provided by Lilian Medina from the KTH in Stockholm. The corresponding results are presented and discussed in Chap. 5.

A.2 Calculation of the Shear Rate in the Microfluidic Channel

The importance of knowing the flow regime in the microfluidic channels has already been discussed in Sect. 2.4. The Reynolds number Re is a straightforward tool used to estimate the flow regime. However, its calculation requires knowing the velocity of the fluid particles—as defined by the Lagrangian description of a flow field—and the viscosity of the fluid. Figure A.6 describes a Poiseuille flow in two dimensions using Cartesian co-ordinates. The fluid flows along the x-axis and its velocity varies as a function of y.

The channel considered here is a square section of side length D_c, so the channel has the cross-section $A_c = D_c^2$. Since it is set by the syringe pumps, the flow rate Q is known and can be used to calculate the average fluid velocity across the section. It holds

$$v = \frac{Q}{h_c w_c}. \tag{A.1}$$

As described in Sect. 7.2, the flow rate can be defined as the velocity difference over a given distance divided by this same distance. In this case, the distance is the height $\frac{h_c}{2}$ and the velocity difference is $v_x(0) - v_x(\frac{h_c}{2})$. However to keep things simple, we will let the velocity difference be approximately equal to the average velocity v. The shear rate in the centre of the channel, where its value is maximal, can thus be expressed as

$$\dot{\gamma} = \frac{2v}{h_c} = \frac{2Q}{h_c^2 w_c}. \tag{A.2}$$

400 μm COC chip - cross-flow The channels in the COC chip are 1 mm wide and 0.8 mm deep (see Fig. 3.4b and c), which means that $h_c = 0.8$ mm and $A_c = 0.8$ mm². Although microfluidic bubbling in Chap. 3 was pressure controlled, one may assume that typical flow rates lie between 100 and 500 μL min⁻¹. Since the chip geometry is cross-flow, the flow rate is divided by two in each channel carrying the liquid before the constriction (see Fig. 3.5), yielding $Q = 50$–250 μL min⁻¹ $= 50$–250 mm³ min⁻¹ $\sim$ 1–4 mm³ s⁻¹. One can thus estimate using Eq. A.2 that typical shear rates in the 400 μm COC chip are in the range $\dot{\gamma} \sim 3$–13 s⁻¹. Looking at Fig. 3.2, one sees that in the microfluidic channels $\eta \sim 0.016$ Pa s. By inserting Eq. A.2 into Eq. 2.20, one can

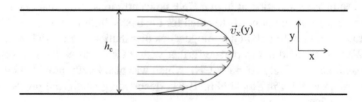

Fig. A.6 Two-dimensional representation of a Poiseuille flow in a channel of height h_c

calculate a range of Reynolds numbers Re over the range of flow rates following

$$Re = \frac{\rho Q}{\eta h_c}, \tag{A.3}$$

which yields, for $\rho \sim 1000$ kg m^{-3}, $Re \sim 0.07$–$0.33 \ll 2000$, confirming that, despite all the approximations made during this calculation, that the flows in the microfluidic chip are laminar (see Sect. 2.4).

190 µm glass chip - cross-flow Let us follow the same reasoning as for the 400 µm COC chip. The channels in the 190 µm glass chip are 0.390 mm wide and 0.190 mm deep (see Fig. 3.4a), which means that $A_c = 0.0741$ mm^2 and $h_c = 0.190$ mm. In Chap. 3, the flows were pressure controlled and one could not determine the flow rates in the chip. To calculate the shear rates in the chip, we have to approximate the flow rates in the microfluidic channel. For the sake of simplicity, let us take the same range of flow rates than for the 400 µm chip, i.e. $Q = 50$–250 µL min^{-1} = 50–250 mm^3 min$^{-1} \sim 1$–4 mm^3 s^{-1} in the channels before the two chitosan flows meet (see Fig. 4.3). One can thus estimate using Eq. A.2 that typical shear rates in the 190 µm chip are in the range $\dot{\gamma} \sim 120$–600 s^{-1}. Looking at Fig. 3.2, one sees that in the microfluidic channels for flow rate between 120 s^{-1} and 600 s^{-1}, the viscosity of the chitosan solution varies little around the value $\eta \sim 0.015$ Pa s, which yields Reynolds numbers in the range of $Re \sim 0.02$–$0.73 \ll 2000$.

In Chap. 4, we used a different chitosan with a lower molecular weight. We used the same 190 µm glass chip but the flow rate was set with a syringe pump and kept constant during microfluidic bubbling with $Q = 180$ µL min^{-1}. The chip geometry being cross-flow, the flow rate is divided by two in each channel carrying the liquid before the constriction (see Fig. 4.3), yielding $Q = 90$ mm^3 min$^{-1} \sim 1.5$ mm^3 s^{-1}. One can thus calculate using Eq. A.2 the shear rate in the 190 µm cross-flow chip: $\dot{\gamma} \sim 213$ s^{-1}. Looking at Fig. 4.2, one sees that in the microfluidic channels $\eta \sim 0.04$ Pa s. Once again, one can calculate the Reynolds number Re in the microfluidic chip using Eq. A.3, which yields, for $\rho \sim 1000$ kg m^{-3}, $Re \sim 0.20 \ll 2000$. The flow of low molecular weight chitosan is thus also laminar in the 190 µm chip with a cross-section geometry.

190 µm glass chip - T-junction The 190 µm glass chip with a T-junction has the same dimensions than its counterpart with a cross-flow geometry, namely $A_c = 0.0741$ mm^2 and $h_c = 0.190$ mm (see Fig. 5.2). However, the flow is not split as in a cross-flow geometry, so that the flow rate is constant with $Q = 180$ mm^3 min$^{-1} \sim 3$ mm^3 s^{-1}. One can thus calculate using Eq. A.2 the shear rate in the 190 µm chip with a T-junction: $\dot{\gamma} \sim 425$ s^{-1}. Looking at Fig. 5.1, one sees that in the microfluidic channels, the viscosity ranges for the different solutions from $\eta \sim 0.04$ Pa s for C40_000 to $\eta \sim 0.06$ Pa s for C41_000. Once again, one can calculate the range of Reynolds numbers in the microfluidic chip using Eq. A.3, which yields, for $\rho \sim 1000$ kg m^{-3}, $Re \sim 0.26$–$0.39 \ll 2000$. The flows are thus all laminar in the 190 µm chip with a T-junction. Let us look closely at the viscosities at the T-junction, where the width of the main channel is halved to 0.195 mm (see Fig. 5.2). The section of the channel becomes A_c

$= 0.0371$ mm^2 and the resulting shear rate is doubled, namely $\dot{\gamma} = 850$ s^{-1}. However, one sees in Fig. 5.1 that the viscosities of the solutions do not significantly vary within the range of shear rates $\dot{\gamma} = 425\text{--}850$ s^{-1}. Therefore, the Reynolds numbers in the constriction of the T-junction do not differ from the Reynolds numbers in the main channel and the flows remain laminar in the constriction.

References

1. Berglund L (2016) Personal communication
2. Medina L (2018) Personal communication
3. Tsianaka A (2016) Synthese und Charakterisierung polydispersen porösen Chitosans hergestellt über Schäumtemplate

Declaration of Authorship

I hereby certify that the dissertation entitled

Monodisperse Highly Ordered and Polydisperse Biobased Solid Foams

is entirely my own work except when otherwise indicated. Passages and ideas from other sources have been clearly indicated.

Name: Sébastien Andrieux

Signed:

Date: 28.02.2018

© Springer Nature Switzerland AG 2019
S. Andrieux, *Monodisperse Highly Ordered and Polydisperse Biobased Solid Foams*,
Springer Theses, https://doi.org/10.1007/978-3-030-27832-8

Curriculum Vitae

Sébastien Andrieux
born 23rd July 1991 in Bergerac, France

Education

09/2009–06/2011	**École Nationale Supérieure de Chimie de Rennes (ENSCR)**, Rennes, France. Undergraduate studies for admission to a French chemistry engineering school.
09/2011–08/2014	**European School of Chemistry, Polymers and Materials Sciences (ECPM), Université de Strasbourg**, Strasbourg, France. *Diplôme d'ingénieur* (specialisation in polymer engineering).
02/2014–07/2014	**Master Thesis, University of Hull**, Hull, United Kingdom, under the supervision of Prof. Bernard P. Binks: *Pickering emulsions stabilised with surfactant crystals formed in situ*.
10/2014–03/2018	**Ph.D. Thesis, University of Stuttgart**, Stuttgart, Germany, under the supervision of Prof. Dr. Cosima Stubenrauch: *Monodisperse Highly Ordered and Polydisperse Biobased Solid Foams*.

Work Experience

07/2011	**FAREVA**, Neuvic sur l'Isle, France. Intern in the quality control laboratory.
05/2013–07/2013	**WACKER CHEMIE AG**, Burghausen, Germany. R&D intern in the *Silicone emulsions and fluids laboratory*.

© Springer Nature Switzerland AG 2019

S. Andrieux, *Monodisperse Highly Ordered and Polydisperse Biobased Solid Foams*, Springer Theses, https://doi.org/10.1007/978-3-030-27832-8

Printed in the United States
By Bookmasters